EMPOWERING TEACHERS THROUGH ENVIRONMENTAL AND SUSTAINABILITY EDUCATION

Empowering Teachers through Environmental and Sustainability Education draws inspiration from an empirical study exploring early career teachers' attempts at enacting Environmental and Sustainability Education (ESE) in their everyday teaching practices. It showcases how a confluence of personal, professional and environmental identities supports implementation of ESE. Additionally, this book discusses key concepts and issues surrounding ESE and the ways in which teachers may claim agency and power to create change in their classroom practices. Drawing from theoretical perspectives, such as Bourdieu's 'thinking tools' habitus and capital, theories of identity, and Foucault's concept of power and knowledge relations, this book explores how teachers negotiate policies, curriculum and institutional norms to further theoretical and practical understanding of ESE. The use of personal narratives offers new insights into teachers' agency in creating localised yet powerful change through small and meaningful actions. The purpose of this book, therefore, is to explore ways in which meaningful change can be made in educational settings through these small agentive and yet empowering steps.

This book reveals that teachers can enact agency and navigate the power structures that exist within educational settings in order to make ESE meaningful within their classrooms.

Melissa Barnes is a Senior Lecturer in the Faculty of Education at Monash University. She works within the fields of teacher education, assessment, policy and TESOL. Melissa has been a classroom teacher in the US, Germany, Vietnam and Australia, collectively shaping her understanding and approach to teaching and learning.

Deborah Moore is a Lecturer in Curriculum & Pedagogy and Early Childhood Education at Deakin University. Deborah's research interests include respectfully researching with young children; listening to children's stories about the places they construct themselves for their own imaginative play; and examining ESE with teachers and learners.

Sylvia Christine Almeida is a Senior Lecturer in Environmental, Sustainability and Science Education in the Faculty of Education at Monash University with strong global teaching experiences across India, Africa, USA, the Middle East and Australia. Her research aims to foreground alternative, non-dominant worldviews in shaping Environmental and Sustainability Education.

EMPOWERING TEACHERS THROUGH ENVIRONMENTAL AND SUSTAINABILITY EDUCATION

Meaningful Change in Educational Settings

Melissa Barnes, Deborah Moore and Sylvia Christine Almeida

Routledge
Taylor & Francis Group
LONDON AND NEW YORK

First published 2021
by Routledge
2 Park Square, Milton Park, Abingdon, Oxon OX14 4RN

and by Routledge
605 Third Avenue, New York, NY 10158

Routledge is an imprint of the Taylor & Francis Group, an informa business

© 2021 Melissa Barnes, Deborah Moore and Sylvia Christine Almeida

British Library Cataloguing-in-Publication Data
A catalogue record for this book is available from the British Library

Library of Congress Cataloging-in-Publication Data
A catalog record has been requested for this book

ISBN: 978-0-367-37039-8 (hbk)
ISBN: 978-0-367-37040-4 (pbk)
ISBN: 978-0-429-35244-7 (ebk)

Typeset in Bembo
by codeMantra

CONTENTS

FIGURES

BOXES

FOREWORD

Dear Readers,

This new text *Empowering teachers through Environmental and Sustainability Education* could not be timelier in many respects. The authors alert to the now precipitous state of the Earth, the recent impacts of extreme weather events including bushfire and the global COVID-19 pandemic. We are currently experiencing a global crisis of a magnitude previously unknown. Churchill is credited with first stating 'never let a good crisis go to waste' and by all accounts this is a 'good' crisis. In every crisis the human populace needs hope for the future, opportunities for critical reflection and different ways to move forward with a sense of agency and empowerment. This publication offers highly pertinent insights from the authors/researchers and the early career teacher study participants to make this happen.

For the early career teacher participants, the overall study intent was to build their sense of hope, agency and empowerment in implementing ESE in their everyday teaching practices. The combination of early career teacher participants, community partners and researchers offers a rich pool of expertise in a conversational and collaborative workshop setting. The research by design approach invites responsiveness to the participants' real-life contexts and stories as the study unfolds. The resulting combined narratives include relatable insights into the professional emergence of early career teachers and the challenges they face in educational settings today around ESE. This book suggests that a somewhat hybrid identity might be the survival strategy needed to fit in as an early career teacher.

Lastly, we live in hope. Despite sustainability now being a cross-curriculum priority in the Australian Curriculum, the ebb and flow of political will around ESE has persisted over decades. The recent youth-led global climate change strikes, ably projected internationally by the media, offer some hope for future generations. Teachers have a key role to play here in promoting hope for regenerating and restoring the Earth. Early career teachers can be change-makers in their educational settings as demonstrated in this text. I strongly encourage all early career teachers, and specifically those involved in this study, to hold on to their ecological identities, continue to push back and please do not waste a 'good' crisis.

Dr Sue Elliott Senior Lecturer and Course
Co-ordinator in Early Childhood Education
University of New England, Armidale, Australia

PREFACE

The three authors—Melissa, Deborah and Sylvia—have been referred to, by others, as being a very 'eclectic group,' with each author having seemingly varied research interests and different educational backgrounds and experiences. Despite this, it was, however, a number of serendipitous events that brought us together, resulting in discussions that prompted the interrogation of the intersections between teacher education, Environmental and Sustainability Education (ESE), teacher identity and the power structures that exist within educational contexts. Our common ground was that we all had the experience of being an early career teacher either in an early childhood setting or a primary or secondary school, trying to manage the multitude of tasks required of us as a new teacher, while trying to strike a balance between accepting the institutional norms of our contexts and aligning our teaching to fit how we understood and related to the world around us. Two of us also had experience working in international settings which added a new layer of cultural and social adjustments to our experiences as teachers. As pre-service teachers, we learnt about *aspirational* educational ideas that had the ability to transform teaching and learning practices, preparing learners for our changing world. However, we found it difficult to enact these ideas within our new classrooms. Without hesitation, we accepted the notion that we were young idealists who had mistakenly thought that we could create meaningful and transformative change within teaching and learning. We accepted that it was our naivete that was the problem, not the educational settings in which we worked.

After many years of teaching in the classroom, we then transitioned to careers in Initial Teacher Education and found ourselves not knowing exactly how to prepare future teachers to create meaningful and transformative change: whether through transformative pedagogies more generally or ESE pedagogies more specifically, in the wide range of educational contexts in which they might find themselves. In this book, we push back against the notion that early career teachers in particular, but teachers in general, must set aside their passions and their naive yet hopeful ideals about creating transformative change in their classrooms. We argue that by navigating and negotiating the educational contexts in which we find ourselves, we can activate our agency by taking small yet agentive and intentional steps towards meaningful implementation of ESE. We also argue that placing value in these small steps supports the emerging teachers' confluence of identities, further building their voice and agency.

Here we re-tell an early discussion between the three of us which signalled the beginning of a supportive and nourishing collaboration and sets the scene for this book:

SYLVIA: The problem is that teachers are just not implementing Environmental and Sustainability Education (ESE).

MELISSA: But I know there are a lot of teachers who are doing things in their schools. I was just talking to a teacher who has built an 'international' veggie patch that allows her English as an Additional Language (EAL) students, alongside their families, to plant and look after a range of vegetables that represent their home countries.

SYLVIA: Ah, yes, the veggie patch. So many teachers think that is what Environmental and Sustainability Education (ESE) is all about. If we have a veggie patch, we must be 'doing' ESE. They think that because it's visible, then they have done it, and don't have to do anything else.

DEB: I know, I've seen that too. In the centre I was researching in recently, there was a veggie patch full of just strawberries, but the children were not allowed to touch them, or dig in the garden or contribute in any way. The Educational Leader at the centre had said that 'only strawberries' were allowed to be grown in there, and I often heard her loudly say, 'get away from there' to any child going anywhere near the 'patch'. Even the water tank didn't work and the children were not given any opportunity to use real tools or water the garden or consider what else could be grown there. I think the 'veggie patch' was just there (next to the see-through fence) for marketing purposes and nothing else. There was a new graduate kindergarten teacher who wanted to grow a whole range of veggies, but became so despondent about the 'strawberries only patch,' she has now left the centre.

SYLVIA: It's as if the veggie patch is a visible, symbolic representation of sustainability for teachers without any depth to their thinking, inquiry, or understanding. It then ends up being a matter of 'ticking the right boxes.'

MELISSA: I know I'm not an expert in sustainability or anything, but we all have to have a starting point, right? What if my first step is putting up a veggie patch in my school? As a teacher, if I had organised to get it going,

FIGURE 0.1   A 'strawberries-only patch' masquerading as a 'veggie patch' purport-
edly for marketing purposes only, positioned amongst the artificial turf
and tanbark, out-of-bounds for children in an early childhood outdoor
setting. (Visual representations by Barnes, Moore & Almeida)

especially knowing that getting buy-in from leadership and colleagues can
sometimes be difficult, I would be really proud of that. Don't we have to
start somewhere?

This excerpt depicts one of many discussions about what ESE implementation
looks like, or *should* look like, in educational settings. Our later discussions re-
volved around the idea that when teachers choose to share, embed and enact ESE
ideas and practices within their teaching and their everyday lives, the decision is
often a very personal one. In addition, teachers' reluctance or willingness to do
so is often grounded in their own experiences, identities and teaching contexts.

Given that Sylvia and Deb had taught in sustainability units in teacher education, Sylvia proposed the idea of a collaborative project that would explore
how teachers, who have strong environmental identities and possess requisite
knowledge of the area, implement sustainability in their teaching. Sylvia had
had a chance encounter, during a teaching placement visit, with an early career teacher who had been one of her former university students and who had
exhibited a strong environmental identity. As a student, she had expressed her
excitement about becoming a teacher so she could utilise the wonderful learning, teaching and research conceptions from university into practice. The conversation quickly veered towards the successes and struggles that this teacher
has experienced in implementing ESE in her new school. It then led to a chain

of other conversations with students from the same Environmental Education unit and who also shared the enablers and barriers in implementing ESE. This conversation prompted us to explore the ways in which teachers are positioned to teach ESE and how their personal, professional and environmental identities are shifted, suppressed and/or elevated in their new teaching roles. This project forced us to extend our understanding of how pre-service teachers engage with the 'idea' of ESE implementation in their future and hypothetical classrooms with their actual experiences of teaching ESE in their new educational settings. This book is an attempt to capture these experiences by re-telling the stories of ten early career teachers as they negotiate educational policies and the politics of teaching and leadership, while also trying to embrace their passion for ESE.

ACKNOWLEDGEMENTS

We would like to thank a number of people who have played a vital role, whether in their support of and participation in the project or the production of this book. First, we want to thank Emeritus Professor John Loughran who awarded us with a Dean's Early Career Researcher grant in 2017, which allowed us to conduct the study on which this book is based. We also thank John for the conversations and for his patient listening to our initial ideas and stirring the research project in a meaningful direction.

Of course, central to this book are those who participated in this project and who have inspired us in a variety of ways, including:

- The 11 early career teachers who invested their time and generously shared their personal, often emotional, stories with us;
- The two participating community partners from the Centre for Education and Research in Environmental Strategies (CERES) and from the Dolphin Institute Research (DRI), whose support, guidance and insight were invaluable to us all;
- The community leaders from sustainability-based organisations in Victoria, such as representatives from Frankston Council, Birdlife Australia, Cool Australia, Melbourne Water, Melbourne Zoo, amongst others, who participated in our project's ThinkTank.

We would also like to thank Colleen Keane and Kathryn Garnier for their expertise in providing proofreading and formatting support, which was generously supported by internal funding provided by both Monash University and Deakin University. Additionally, we would like to thank the reviewers of this book who provided targeted feedback and advice on positioning our Australian study within a broader international context. Finally, we want to thank our families for supporting us and giving us the time and space to make this book possible.

We anticipate that as you read the stories of these early career teachers, which are re-narrated throughout this book, their experiences will resonate with your own stories; yet you will be filled with hopeful aspirations that small, intentional steps can add value to the larger cause and lead to transformative and meaningful change for the future.

1

ESE

Finding hope amidst future concerns

Introduction

Environmental concerns, issues and conversations are a key part of our everyday conversations. We are constantly being asked to take sides—either as supporters or deniers.

Climate emergency has undoubtedly been a contentious issue that has united or divided public opinion. Eleven thousand scientists declared that climate change conditions are now severe enough for us to rename this a 'climate emergency' demanding urgent action (ABC, 2019a). A key figure that has helped rally public opinion and evoked activism especially for the young is Greta Thunberg. The photo below (Figure 1.1) depicts a poster of Greta Thunberg that was placed on the noticeboard within our faculty as a sign of support for her movement. The anonymous graffiti on this poster provides a clear and diametrically opposing view to Greta and her influencers, thereby suggesting that young people cannot and should not have a voice. Their power and voice only comes from adults surrounding them. This in many ways is a powerful reminder to many of us about the role of media, scepticism and the many forces we as educators battle in our professions.

This book focuses on the constant tussles that passionate and motivated teachers with strong environmental identities grapple within their everyday practice. The book highlights the main issues, the power of education in providing a voice to young children and the journeys of teachers who are keen to make a difference.

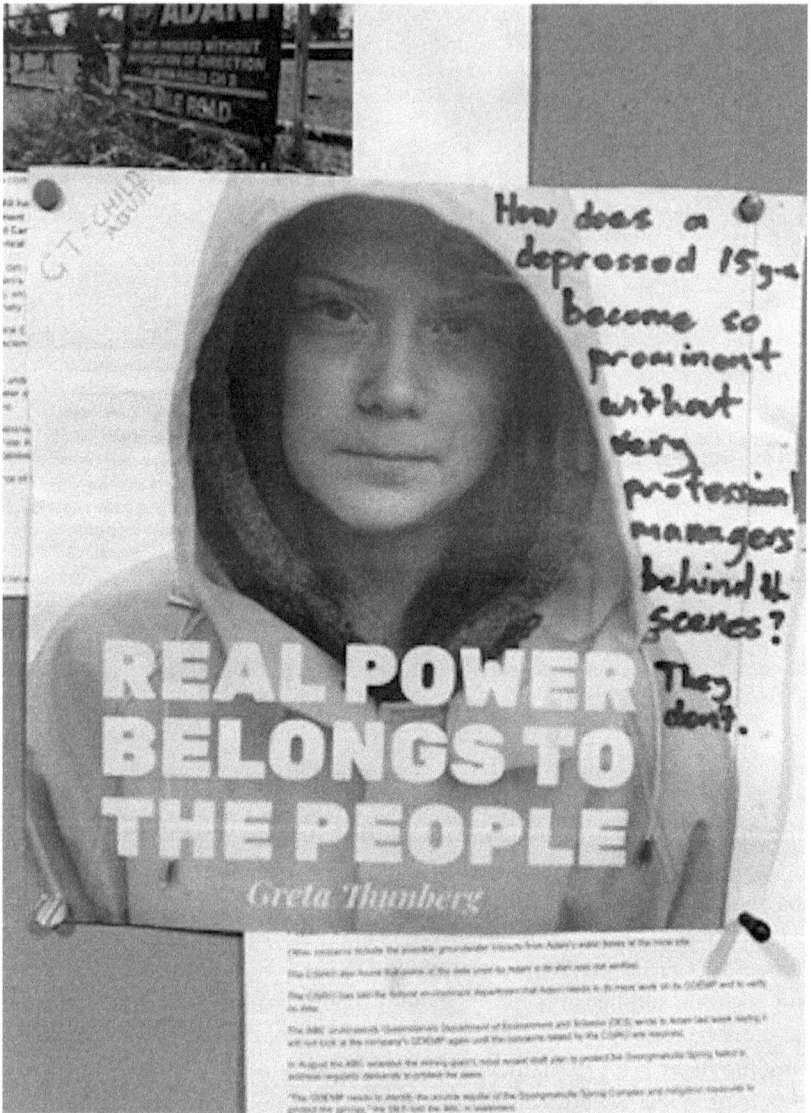

FIGURE 1.1 Poster of Greta Thunberg that had been posted on a noticeboard at the university. (Visual representations by Barnes, Moore & Almeida)

Global environmental challenges

Environmental challenges have seen an unprecedented increase ranging from global population explosion (approximately 7.9 billion), higher rates of urbanisation, intrusive technological interventions like geo-engineering and nuclear warfare, slower economy fuelling trade wars, rising inequity caused by conflicts resulting in large-scale migration and triggering the refugee crisis and, ultimately, the overarching issue of climate change. The United Nations Environment Protection (UNEP) agency's recent comprehensive Global Environment Outlook Report 6 (UNEP, 2019) identifies five major drivers for environmental degradation, namely population growth and demographics, urbanisation, economic development, new technological forces and climate change. It reckons these have created imbalance in wealth and access to quality healthcare, thereby offsetting gains in life expectancy and quality of life. They impact critical issues ranging from resource depletion, biodiversity loss, water scarcity, health, pollution and environmental degradation.

The Intergovernmental Science-Policy Platform on Biodiversity and Ecosystem Services (IPBES, 2019) in its first landmark report compiled by 145 experts from 50 countries highlights that nearly one million species are threatened with extinction and that global response has been insufficient in bringing about 'transformative change' that supports fundamental, system-wide reforms through technological, economic and social changes. Sir Robert Watson, its chair, points out that 'the health of ecosystems on which we and all other species depend is deteriorating more rapidly than ever. We are eroding the very foundation of our economies, livelihoods, food security, health and quality of life worldwide' (p. 1).

Notable findings of the report point out that 75% of land and 66% of marine environments have been altered by humans with more than one-third and nearly 75% of freshwater devoted to growing food. Findings also show the excessive unsustainable fishing patterns and plastic pollution of the oceans alongside the near doubling of urban areas in only 25 years (IPBES, 2019). The climate change crisis continues to dominate with increased concerns of permanent damage to environments and loss of life caused by erratic weather patterns. The latest Intergovernmental Panel on Climate Change report emphasises that climate change is directly impacting people, ecosystems and livelihoods on global scales (IPCC, 2019).

Two major events have preceded the writing of this book. The first major event is the COVID-19 pandemic which continues to engulf the world, straining health systems and economies across the world. Most nations have shown its long-lasting impacts on population and economy. This has brought attention to the frail human environment relationship and the delicate balance that supports the interdependence of human health on healthy ecosystems (Armstrong, Capon & McFarlane, 2020). COVID-19 outbreak was a direct result of a lack of respect for nature and biodiversity, thereby causing a jump in disease-causing viruses from these non-domesticated animals to humans. Given our human systems have not encountered these viruses before, the impact is devastating, as witnessed by the rising number of cases and related deaths. By mid-November 2020, there have been 54,296,615 globally confirmed cases with 1,315,881 deaths, providing a glimpse of the massive impact this pandemic is having across the world with numbers rising exponentially (John Hopkins, 2020).

The second major event has been the recent bush fires that devastated Australia in early 2020. These bush fires were a prudent reminder of local environmental impacts directly related to climate change during one of the hottest summers on record in Australia. The bush fires this summer in Australia have burned through ten million hectares of land, killed dozens of people and more than half a billion wildlife, destroying large communities and habitats. Australia has seen its hottest year on record and its most destructive bush fire season (news.com, 2020). The year 2019 alone has seen three major fires burning large swathes of the Amazon jungle, the Siberian wilderness and now the Australian bush. While large fires have been seen at other times in history, what is unprecedented is the size, intensity and frequency—all of which point clearly towards climate emergency (Guardian, 2020).

Climate change leads to habitat loss and shifting climate patterns, thereby causing migration of wildlife moving to new places where they might cause novel diseases through their encounters with new species. The close connection between the pandemic and climate change serves as a reminder to take into account 'planetary consciousness' and strive towards maintaining respect for the natural environment (Armstrong et al., 2020). Covid-19 has certainly opened everyone's eyes to the many possibilities that can also positively impact the climate agenda. This includes reducing carbon emissions by working from home, reduced pollution and green transitions to support economic growth. One example is the river Yamuna, in India, which showed a marked difference in the level of pollution before the pandemic and two months into the lockdown in Delhi (Insider, 2020). There have been multiple attempts and monetary investment to the tune of millions of dollars to clean up the river over more than 25 years now. The pandemic and reduced pollution have led to a natural clean-up in less than three months.

Tackling the pandemic beyond a human/economic perspective and addressing the underlying issues of biodiversity conservation and climate stabilisation are important and provide a holistic approach for future responses.

What is environmental and sustainability education? Where does education fit into this?

Environmental Education (EE), Education for Sustainable Development (WCED, 1987, p. 43), Education for Sustainability (EfS), Environmental Education for Sustainability (EEfS) and Environmental and Sustainability Education (ESE) are terms that are often used interchangeably in relation to sustainability and education. A quick look at the historical perspectives and emergence of these terms here will help to understand the journey and trajectories of the field.

Before offering a review, it is imperative to point out the ambivalence and debate around the definitions and/or terminology. One of the key barriers in the uptake of sustainability has been this lack of strongly defined terminology that takes into account the multiple perspectives of an ever-changing global situation. A clearer standpoint is urgently required, given that the most widely used definition, namely the Brundtland definition created by the World Commission on

Environment and Development (WCED) in 1987, cannot hold up today, given the many new changes in global systems (WCED, 1987).

Critiques have viewed sustainable development either as an oxymoron of irreconcilable opposites, or a paradox of contradicting terms or a still emerging concept that is yet to take a firm shape (Brown, Grootjans, Ritchie, Townsend & Verrinder, 2005). This definition is contradictory, pitching ecological and economic development as opposites, when there is ample evidence that shows that the only way forward is to take both into account.

Brief history of environmental and sustainability education

According to the existing literature, the first documented use of *EE* in an international setting was at a meeting of International Union for Conservation of Nature (IUCN) in Paris in 1948 (Disinger, 1982). Later, in 1970, IUCN held a conference themed *Environmental Education in the School Curriculum* in Nevada, USA, where Palmer and Neal (1994) claim that deliberations from that meeting have influenced major developments in EE, where IUCN officially defined EE in 1970:

> Environmental education is the process of recognising values and clarifying concept in order to develop skills and attitudes necessary to understand and appreciate the interrelatedness among man, his culture and his biophysical surroundings. Environmental education also entails practice in decision-making and self-formulation of a code of behaviour about issues concerning environmental quality.
>
> *(IUCN, 1970, p. 65)*

In 1980, with the unveiling of the concept of *Sustainable Development*, experts highlighted that environmental conservation and development should be mutually interdependent and this consensus led to another key international conference on environment and development *The Earth Summit*, which was held in Rio de Janeiro in 1992. As a key outcome of this summit, experts recommended bringing both environmental and sustainable development education together, within both formal and non-formal education sectors, as a crucial part of learning (Palmer & Neal, 1994). Another major breakthrough of the summit was the introduction of the *Rio Declaration*, which set out 27 principles for environmental protection and responsible development, reaffirming the declaration of the UN conference on the Human Environment adopted at Stockholm on 16 June 1972 (Palmer & Neal, 1994). Among them, the tenth principle was:

> Environmental issues are best handled with the participation of all concerned citizens, at the relevant level. At the national level, each individual shall have appropriate access to information concerning the environment that is held by public authorities, including information on hazardous

materials and activities in their communities, and the opportunity to participate in decision-making processes. States shall facilitate and encourage public awareness and participation by making information widely available. Effective access to judicial and administrative proceedings, including redress and remedy, shall be provided.

(United Nations, 1992)

In view of that, one motive of the declaration was to increase the environmental sustainability awareness through information and education. Since 2000, agencies of the UN worked together with worldwide governments to focus on Sustainability Education (SE), and among many of them there are four key initiatives which focus solely on EfS in one way or another (O'Flaherty & Liddy, 2017). These are:

- The Millennium Development Goals (MDGs)
- Education for All (EFA)
- The United Nations Literacy Decade (UNLD)
- The United Nations Decade of Education for Sustainable Development (UNDESD)
- Decade of Action 2020–2030

Among them UNDESD's primary intention was to promote ESE through making people aware that education is an important pathway to a sustainable future. However, it is apparent that merely raising awareness is inadequate to create change, where ESE should intensely encourage the need for personal initiatives and public participation in order to achieve a sustainable way of living (Lambrechts, Liedekerke & Petegem, 2017; O'Flaherty & Liddy, 2017; Palmer, 1998; Palmer & Neal, 1994).

In 2014, after years of concerted effort and conversation, the Sustainable Development Goals (SDGs) were adopted by most nations during the UN SDG Summit (UN, 2015), with the intention of creating a better and more sustainable future for everyone. The 17 SDGs are interconnected, providing an all-encompassing global effort to solve some of the most critical issues impacting our planet. The SDGs are expected to play a central role in policy and planning documents across the globe and are deemed to be achieved by 2030. In lieu of the short time span available to achieve these far-ranging goals, the UN declared 2020–2030 as the Decade of Action (UN, 2019).

Goal 4 of the SDGs is Quality Education (UN, 2019) aimed at upward socio-economic mobility and an opportunity to escape the cycle of poverty. This goal also aims to reform the lives of many young girls who are denied any education (let alone quality education), often due to social roles and environmental factors like having to walk longer distances in search of water and firewood for cooking. Many countries are working towards achieving the SDGs, especially by encouraging education for all through major policy developments. A case in point is India and the range of recent education initiatives, like the recently released National Education Policy (MHRD, 2020), which particularly aims at fulfilling Goal 4 for all children. Other large governmental initiatives like the Sarva Shiksha Abhiyan are aimed at providing universal access to education for all primary and middle school students in India (AICTE, 2020) with an outreach to over 192 million children. This emphasis on meeting Goal 4 is also seen in the early childhood policies with the new national curriculum asking for uniform educational opportunities for all young children while keeping the socio-economic barriers, caste, religion, gender and other disparities in mind (Almeida & Ohara, 2020).

Our choice of terminology

Education has become an indispensable instrument for accomplishing sustainability over recent years, as highlighted in the earlier section. The SDGs provide a prominent role for education promoting it as central to all initiatives taken for generating a deeper, long-lasting impact. Many educators and policymakers have used the term *Sustainable Education, Education for Sustainable Development and Education for Sustainability* interchangeably instead of the term *Environment and Sustainability Education* (Laurie, Nonoyama-Tarumi, Mckeown & Hopkins, 2016; O'Flaherty & Liddy, 2017; Wals & Benavot, 2017).

We use ESE as our chosen terminology for this book for a range of reasons. One of the key reasons for using ESE is that it brings Environment to the foreground and links it to Sustainability without allowing the prepositions like 'of,' 'in,' and 'for' to shift the focus in any way. It also keeps the term 'Development' out, thereby allowing for a holistic rather than a resource-driven approach. The broadest meaning of ESE is that it is about developing a sense of ownership, empowerment and building capacity among communities to address environmental and sustainability issues in their own neighbourhoods through their everyday practices (Wali, Alvira, Tallman, Ravikumar & Macedo, 2017). Accordingly, ESE is about moving in tune with people's beliefs, attitudes, so that their deeper conscience will want to act sustainably while including a sense of solidarity (Atkinson & Wade, 2014). Palmer and Birch (2003) further clarify that ESE should be able to help people to translate their values and attitudes into action through knowledge. This allows for informed decisions individually, as well as collectively, for the well-being of the society. Environmental sustainability experts and scholars believe that the collective aims of ESE should focus on developing the critical awareness of ecological, social, economic and political components, in order to work towards a quality life and a sustainable future (Palmer & Birch, 2003; Vare & Scott, 2007; Wali et al., 2017). More importantly, ESE delivers the message of interdependent nature among living systems and the consequences of individual and collective human actions, both now and in the future (Bonnett, 2006; Palmer, 1998; Palmer & Birch, 2003). We also acknowledge the debates around terminologies from EE, ESD, EfS and EEfS. For the purpose of this book, we choose to use the term ESE, as it uses both terminologies of 'environment' and 'sustainability' equally, without prioritising one over the other. We are conscious of using this as a synonym to the other terminologies presented earlier, but aim to use ESE as the overarching umbrella term that includes all the others.

Overall, education can be effective in creating awareness of the ramifications caused by human behaviours and increasing the sensitivity towards environmental issues (Mohiuddin, Mamun, Syed, Masud & Su, 2018). More importantly, both formal and non-formal education are decisive pathways to accomplish environmental and ethical consciousness, encouraging values, attitudes, skills and behaviours that aid sustainable development concerns and effective public participation in decision-making (Fuertes-Camacho, Graell-Martín, Fuentes-Loss & Balaguer-Fàbregas, 2019; Mohiuddin et al., 2018; Wals & Benavot, 2017). Therefore, it is vital that governments focus on enhancing the availability and the accessibility of ESE to the public and ESE should be a part of lifelong learning processes (European Commission, 1997). Scholars highlight that modern ESE encompasses more than just manoeuvres; and they suggest that ESE must build upon research, outreach and integrated curricula which enable holistic and penetrative perspectives on educational outcomes (Koehn & Uitto, 2013; Martin & Samels, 2012). Hence, education experts, researchers and practitioners must evaluate the contemporary demands of ESE in both formal and non-formal education settings to increase the adoptability and positive impacts of ESE.

Why should you read this book

This book provides insight into early career teachers' experiences in implementing ESE in their initial years of teaching. It appeals to pre-service training preparing them to understand some of the key issues, barriers and supports as they follow their passion into that first teaching position post their graduation. In service, teachers will find synergies with their own experiences when they start their teaching careers. Many of these teachers are now mentoring the newly graduated teachers in their schools—this book will share some of the joys, pains and struggles of their mentees and help them understand their journeys better. For policymakers, the book provides an insight into teachers' experiences in putting into practice the policies that direct and dominate their teaching. Policies, when not supported by strong dissemination and uptake, end up being aspirin-level solutions that offer little more than lip-service to the existing issues of reform (Almeida, 2015). This book will showcase which aspects of policy uptake are successful and where/how early career teachers can be better supported to implement ESE policies in their new classrooms. For readers interested in the scholarship around theories, this book extends and pushes theoretical boundaries by offering new ways of theorising teachers' experiences using the confluence of identities and Bourdieu's *habitus*. It also offers new perspectives in narrative inquiry combined with research by design as a methodology, offering a distinctive three-dimensional narrative inquiry approach as a lens of analysis on temporal, societal and place-based dimensions. These theoretical and methodological elaborations should also appeal to researchers and students pursuing research studies as they provide practical glimpses into the ways to connect theory, policy and practice. Our hope in writing this book has been to reach out to a large audience ensuring it caters to the interests and curiosities of anyone interested in contemporary education.

References

ABC. (2019a). Climate change emergency. Retrieved on January 6 from https://www.abc.net.au/news/2019-11-06/climate-change-emergency-11000-scientists-sign-petition/11672776

ABC. (2019b). Fires misinformation being spread through social media. Retrieved on January 9 from https://www.abc.net.au/news/2020-01-08/fires-misinformation-being-spread-through-social-media/11846434?section=science

AlCTE. (2020). Sarva Shiksha Abhiyan. Retrieved on October 13, 2020 from https://www.aicte-india.org/reports/overview/Sarva-Shiksha-Abhiyan

Almeida, S. C. (2015). *Environmental education in a climate of reform*. Rotterdam, The Netherlands: Sense Publishers.

Almeida, S. C., & Ohara, Y. (2020). Research in early childhood education for sustainability: Policies and perspectives from India. In S. Elliott, E. Ärlemalm-Hagsér & J. Davis (Eds.), *Researching early childhood education for sustainability: Challenges, assumptions and orthodoxies* (pp. 82–93). New York, NY: Routledge.

Armstrong, F., Capon, A., & McFarlane, R. (2020). Coronavirus is a wake-up call: Our war with the environment is leading to pandemics. Retrieved June 01, 2020 from https://theconversation.com/coronavirus-is-a-wake-up-call-our-war-with-the-environment-is-leading-to-pandemics-135023

Atkinson, H., & Wade, R. (2014). *The challenge of sustainability: Linking politics, education and learning*. Bristol, UK: Policy Press.

Bonnett, M. (2006). Education for sustainability as a frame of mind. *Environmental Education Research, 12*(3–4), 265–276. doi:10.1080/13504620120109619

Brown, V., Grootjans, J., Ritchie, J., Townsend, M., & Verrinder, G. (2005). *Sustainability health: Supporting global ecological integrity in public health*. Crows Nest, Australia: Allen & Unwin.

Disinger, J. F. (1982). Environmental education research notes (USA). *Environment Systems and Decisions, 2*(3), 203–206. doi:10.1007/BF02603098

European Commission. (1997). Paper presented at the Environmental Education and Training in Europe, Brussels, Belgium.

Fuertes-Camacho, M. T., Graell-Martín, M., Fuentes-Loss, M., & Balaguer-Fàbregas, M. C. (2019). Integrating sustainability into higher education curricula through the project method, a global learning strategy. *Sustainability, 11*(3), 767–792. doi:10.3390/su11030767

Guardian. (2020). Retrieved on January 9 from https://www.theguardian.com/australia-news/2019/nov/22/australia-bushfires-factcheck-are-this-years-fires-unprecedented

Insider.com. (2020). Before-and-after photos show the dramatic effect lockdowns are having on pollution around the world. Retrieved on June 15, 2020 from https://www.insider.com/before-after-photos-show-less-air-pollution-during-pandemic-lockdown

Intergovernmental Panel for Climate Change. (2019). Global warming of 1.5°C. Retrieved on February 26, 2020 from https://www.ipcc.ch/site/assets/uploads/sites/2/2019/06/SR15_Full_Report_High_Res.pdf

Intergovernmental Science-Policy Platform on Biodiversity and Ecosystem Services. (2019). Retrieved on February 26, 2020 from https://www.un.org/sustainabledevelopment/blog/2019/05/nature-decline-unprecedented-report/

IUCN. (1970). International working meeting on environmental education in the school curriculum. Retrieved from https://portals.iucn.org/library/node/10447

John Hopkins. (2020). Coronavirus resource centre tracking. Retrieved on November 16, 2020 from https://coronavirus.jhu.edu/

Koehn, P. H., & Uitto, J. I. (2013). Evaluating sustainability education: Lessons from international development experience. *International Journal of Higher Education Research, 67,* 621–635. doi:10.1007/s10734-013-9669-x

Lambrechts, W., Liedekerke, L. V., & Petegem, P. V. (2017). Higher education for sustainable development in Flanders: Balancing between normative and transformative approaches. *Environmental Education Research, 24*(9), 1284–1300. doi:10.1080/13504622.2017.1378622

Laurie, R., Nonoyama-Tarumi, Y., Mckeown, R., & Hopkins, C. (2016). Contributions of education for Sustainable Development (ESD) to quality education: A synthesis of research. *Journal of Education for Sustainable Development, 10*(2), 226–242. doi:10.1177/0973408216661442

Martin, J., & Samels, J. E. (2012). *The sustainability university: Green goals and new challengers for higher education leaders.* Baltimore, MD: Johns Hopkins University Press.

Ministry of Human Resources Development (MHRD). (2020). National Education Policy. Retrieved on October 13, 2020 from https://www.mhrd.gov.in/sites/upload_files/mhrd/files/NEP_Final_English_0.pdf

Mohiuddin, M., Mamun, A. A., Syed, F. A., Masud, M. M., & Su, Z. (2018). Environmental knowledge, awareness, and business school students' intentions to purchase green vehicles in emerging countries. *Sustainability, 10*(5), 1534–1552. doi:10.3390/su10051534

news.com.au. (2020). Retrieved on January 9, 2020 from https://www.news.com.au/technology/environment/how-the-2019-australian-bushfire-season-compares-to-other-fire-disasters/news-story/7924ce9c58b5d2f435d0ed73ffe34174

O'Flaherty, J., & Liddy, M. (2017). The impact of development education and education for sustainable development interventions: A synthesis of the research. *Environmental Education Research, 24*(7), 1031–1049. doi:10.1080/13504622.2017.1392484

Palmer, J. A. (1998). *Environmental education in the 21st century: Theory, practice, progress and promise.* London, UK: Routledge.

Palmer, J. A., & Birch, J. C. (2003). Education for sustainability: The contribution and potential of a non-governmental organisation. *Journal of Environmental Education Research, 9*(4), 447–468. doi:10.1080/1350462032000126104

Palmer, J. A., & Neal, P. (1994). *The handbook of environmental education.* Chatham, UK: Mackays of Chatham.

sUNEP. (2019). Retrieved February 26, 2020 from https://content.yudu.com/web/2y3n2/0A2y3n3/GEO6/html/index.html?page=56&origin=reader

United Nations. (1992). The Rio declaration on environment and development [Press release]. Retrieved from http://www.unesco.org/education/pdf/RIO_E.PDF

United Nations. (2015). The sustainable development goals. Retrieved on June 1 2020 from https://www.un.org/sustainabledevelopment/sustainable-development-goals/

United Nations. (2019). Decade of action. Retrieved on June 1, 2020 from https://www.un.org/sustainabledevelopment/decade-of-action/

Vare, P., & Scott, W. (2007). Learning for a change: Exploring the relationship between education and sustainable development. *Journal of Education for Sustainable Development, 1*(2), 191–198. doi:10.1177/097340820700100209

Wali, A., Alvira, D., Tallman, P. S., Ravikumar, A., & Macedo, M. O. (2017). A new approach to conservation: Using community empowerment for sustainable well-being. *Ecology and Society, 22*(4), 70–83. doi:10.5751/ES-09598-220406

Wals, A. E. J., & Benavot, A. (2017). Can we meet the sustainability challenges? The role of education and lifelong learning. *European Journal of Education, 52*(4), 404–413. doi:10.1111/ejed.12250

World Commission on Environment and Development. (1987). *Our common future.* Oxford: Oxford University Press.

2

EDUCATIONAL LANDSCAPES

ESE curricular initiatives and change

While environmental education has made its way into national curricular policies and guidelines in a number of countries (ACARA, 2020a; Lee & Kim, 2017), these *aspirational* policies have not always translated into sustainable curricular changes or practices within educational contexts (Almeida, 2015; Munoz-Pedreros, 2014). There are numerous Environmental and Sustainability Education (ESE) initiatives evident within curriculum globally and across education contexts; however, many argue there has been a limited focus on documenting, analysing and reporting these initiatives (Parker, 2017; Sprenger & Nienaber, 2018; White, Eberstein, & Scott 2018; Yi Lo, 2010) and that the road to meaningful curricular change is slow (Munoz-Pedreros, 2014; Li, 2013).

This chapter will examine ESE curriculum initiatives and change within the literature, highlighting change from a variety of educational contexts in a range of different countries. It will conclude with a focus on a curricular initiative implemented in one state/territory in Australia as a way to examine how infrastructure and resources are important for curricular change. It also highlights how creating curricular change is complex and slow and there is a number of small but critical steps that are required in order for change to be substantive in Australian schools and early childhood centres.

ESE curricular change: higher education

The purpose of this literature review is to focus primarily on ESE curriculum initiatives and approaches in both early childhood and schooling (K-12) contexts. However, the existing literature on ESE curricular initiatives reveals an emphasis on how universities and other higher education institutions implement the Sustainable Development Goals (UNESCO, 2017), more specifically Education for Sustainable Development (Barth & Rieckmann, 2016; Sprenger & Nienaber,

2018) and, therefore, it would be remiss to exclude higher education contexts from this discussion.

The role of ESE as 'a problem-solving enterprise' (Karim, Srisandarajah & Heiter, 2013, p. 76) is central to a number of studies that have examined how to integrate or infuse ESE into higher education curricula (Karim et al., 2013; Rowe & Johnston, 2013; Tejedor, Segalàs, Rosas-Casals, 2018; Vincent & Focht, 2011). Tejedor et al. (2018) argue that sustainability has transdisciplinary aspects, incorporating perspectives from social sciences and humanities which are unfamiliar to learners in higher education disciplines, such as engineering. However, they found that to encourage novel ways of knowledge production in higher education contexts, a transdisciplinary approach encourages a form of reasoning that is more fluid than problem-solving in most science disciplines and can help faculties overcome disintegration among disciplines.

Almeida (2015), in her doctoral thesis, studied the uptake of ESE among teacher educators in India, which was found lacking despite strong policy reforms. Key barriers to the implementation of ESE were lack of professional development, poor resources, crowded curriculum, limited pedagogical skills and ineffective policy dissemination. The study found a mismatch between personal and professional identities with this group of teacher educators unable to drive their strong environmental sensibilities to embed within their teaching practices. In another study, Li (2013) explored how ESE was understood and taught by teachers within University English programmes from six provinces in China. Li (2013) argues that fusing English language learning, a skill needed for communicating in a global world with content that exposes learners to global issues and social responsibility, can be mutually educative. However, she argues that teachers' desire and capacity to teach ESE is hindered by a lack of teacher training, a narrow curriculum and slow-moving administration and policy. Likewise, Barth and Rieckmann (2012) argue that for university curricula change to be truly transformed, with sustainability built into the curriculum rather than superficially applied, there needs to be a focus on developing the academic staff as transformative change relies on their willingness and capability to support the process.

In the US context, Powers (2004) contends that introducing ESE into the curriculum of teacher education programmes, based within higher education institutions, has a 'multiplier effect' (p. 3), as the impact goes beyond future teachers but extends to their future learners. However, at the time of her report, she argued there had been limited impetus to introduce ESE into teacher education programmes—regardless of the impact it might have. Powers (2004) further argues that many schoolteachers are willing to incorporate ESE but require professional training, which she claims should happen during their university studies.

ESE in early childhood curricula

Like higher education contexts, which are characterised as having curricula flexibility (Vincent & Focht, 2011), early childhood learning frameworks and curricula provide a similar flexibility and a viable space for ESE integration. For example, the Early Years Learning Framework (EYLF) in Australia is:

> a framework for interpretation… It offers a unique opportunity for early childhood professionals to engage in critically reflective decision-making about how the themes, principles and outcome of EYLF are to be interpreted.
>
> *(Elliott, 2014, p. 7)*

The EYLF is framed by the concepts Belonging, Being and Becoming—all of which are framed by a focus on relationships with others. Initially, the Australian EYLF (2009) and National Quality Framework (NQF) (2012) required Australian early childhood educators to embed 'sustainable practices' as evident in NQF: Standard 3, and EYLF: Outcome 2, as follows:

National Quality Standard 3 (NQS)
- **Element 3.3.1 Sustainable practices are embedded in service operations**:
  Educators and children work together to learn about and promote the sustainable use of resources and to develop and implement sustainable practices.
- **Element 3.3.2 Children are supported to become environmentally responsible and show respect for the environment**:
  Children develop an understanding and respect for the natural environment and the interdependence between people, plants, animals and the land.

Early Years Learning Framework—Outcome 2
- *Children are connected with and contribute to their world*
  Children become socially responsible and show respect for the environment.
  Educators promote this when they embed sustainability in daily routines and practices and when they find ways of enabling children to care for and learn about the land.

Importantly, Elliott (2014) has long argued that the EYLF and NQF approach to sustainability, as seen here in these policy documents, is fundamentally anthropocentric—with humankind being central to a child's belonging, being and becoming without acknowledging how animals and the natural environment play a key role. She laments, 'It is this prioritising of humans, their relationships and needs without consideration for our interdependent relationships with the Earth that has led to the current global concerns' (p. 7). With a focus on sustainable futures, there is an urgent need to reinterpret the EYLF in light of children's relationships with the Earth. While the NQF standards and EYLF outcomes presented here show that 'sustainability' was frequently mentioned, it does not make it clear how this could be done through the 'weak' terminology used to define what 'sustainable practices' could look like (Elliott & Davis, 2018, p. 172). Furthermore, Elliott and Davis (2018) argue that the EYLF at its inception appeared to focus more on 'promoting outdoor play in natural settings almost in lieu of more multi-dimensional EC EfS approaches' (p. 166), suggesting that the inclusion of sustainability was rhetorical rather than a real expectation of teachers in early childhood settings. It suggests that outdoor play (or nature play) is enough in itself to promote EC ESE.

Of particular relevance to this book, the National Educational Leader and General Manager of the Australian Children's Education and Care Quality Authority (ACECQA), Rhonda Livingstone (2016), commented:

> You are not alone if you find yourself challenged when thinking about 'embedding' sustainability into your service, or how to engage young children in learning about environmental responsibility....

Although articles on how to enact sustainable practices were available on the ACECQA website, Livingstone's (2016) statement is indicative of the underlying assumption that it was 'too challenging' to try to embed or enact sustainability in early childhood settings. This attitude was reinforced when in 2018, NQS 3.3 was completely removed from the NQF requirements, apparently due to it 'proving too difficult for educators to implement and [therefore] should not be part of government-legislated quality criteria' (Elliott & Davis, 2018, p. 172). Alarmingly, at this time, the only policy that now relates to sustainability in the NQF is as follows:

National Quality Standard 3 (NQS)
• Element 3.2.3: Environmentally responsible
    The service cares for the environment and supports children to become environmentally responsible

While Elliott and Davis (2018) consider this dramatic shift in policy has a 'clear intent... to weaken the already rather weak references to sustainability across the entire education spectrum' (p. 172), they hope that the increasing number of early childhood educators, 'networks and professional associations for the inclusion of EfS, [are] strong enough to withstand such a watering down of policy' (Elliott & Davis, 2018, p. 172).

Australia is not the only country that possesses curricular frameworks and/or guidelines in the early childhood space. While some early childhood curricular documents position ESE as an important feature of the curriculum, others are ripe with opportunities to further interpret their early childhood frameworks, ensuring that ESE is positioned as an empowering tool rather than as a tokenistic gesture. In both Korea and Sweden, ESE is a key component of their national early childhood curricula frameworks (Ärlemalm-Hagsér & Engdahl, 2015; Davis, 2015; Ji, 2015). In addition, Hong Kong has curricular guidelines for how to conduct ESE in early childhood (Yi Lo, 2010) and Japan's early childhood guidelines include 'nature-based' objectives (Inoue, 2015). However, regardless of the placement of ESE in curricular framework and guidelines, or the reinterpretation of ESE concepts into existing frameworks, as in the case with Australia's EYLF, challenges still exist with uptake of ESE and the professional learning of teachers. While macro-policy approaches (e.g. national ESE curricula) is a positive way forward, for change to occur there also need to be grassroots approaches or a micro-policy approach that empowers and practically equips teachers to enact these ESE curricular initiatives. For example, Ji (2015) argues that teachers need practical ways in which to implement ESE, particularly in how they connect with their parents and the wider community, and how to engage with other environmental professionals. Similarly, Ärlemalm-Hagsér and Engdahl (2015) suggest that it is the diversity of attitudes, values and behaviours among early childhood teachers that can make it difficult for some teachers in Sweden to implement ESE in an empowering way.

Similarly, Yi Lo (2010) argues that while there has been a policy-practice gap in Hong Kong in the past, with many early childhood teachers unsure of how to translate the ESE guidelines into practice, there has been a positive change in ESE uptake with the help of teaching resources produced by the government and other organisations. However, she suggests that the overcrowded curriculum and vague assessment tasks make it even more difficult for teachers who lack knowledge and/or the self-efficacy to teach and implement ESE activities.

Researchers argue that ESE plays an important role in a young child's life because natural and living environments offer endless opportunities to discover and learn (Inoue, 2015; Wilson, 2012; Yi Lo, 2010; Young & Elliott, 2003) and children possess '*basic rights* to survival, protection and participation' (Davis, 2015, p. 25). In other words, ESE allows an opportunity for young children to learn, interact and understand their natural environment while also taking part in the decision-making and actions that are required for sustainable living. However, Elliot, Ärlemalm-Hagsér and Davis (2020) argue that caution should be taken to not equate ESE with nature play. While children should be encouraged to interact with nature and the spaces around them (see more about place-based learning in Chapter 6), this needs to be complemented by transformative and *intentional* learning about ESE. In order to empower young children to learn and interact in their natural environments in meaningful and intentional ways, the literature suggests that we must close the policy-practice gap. By focusing on children's agency to engage with ESE, Weldemariam and Wals (2020) suggest that this prevents early childhood educators from 'rethinking the deep-rooted anthropocentric assumptions and practices that dominate' ESE in early childhood settings. While there is a multitude of resources that teachers can use to implement ESE in an early childhood environment, the question still remains, as articulated by Ärlemalm-Hagsér and Engdahl (2015, p. 261): 'Why don't more teachers engage in empowering forms of EfS in preschools...?' Therefore, there needs to be a focus on disrupting the beliefs and practices of educators and of supporting their professional as well as environmental identities in order to empower children in early childhood settings.

ESE in primary and secondary curricula

While ESE is deemed an important topic to include in primary and secondary curricula, given that young people must rely on the environment to provide the resources to sustain their lives (Littledyke, Taylor & Eames, 2009), the literature highlights the complexity of attempting to implement ESE in a meaningful and sustainable way. First, there are arguments made in the literature with regard to indoor/outdoor opportunities for learning and the importance placed on 'formal' learning. This section will highlight how several researchers and educators have attempted to make sense of ESE as a classroom-based, inquiry-driven and problem-solving learning opportunity for learners. Second, the literature, to date, discusses how ESE has made its way into national curricula and developed 'model' schools, highlighting the role of government initiatives in ESE implementation in primary and secondary schools. Finally, similar challenges, which have plagued both higher education and early childhood contexts, are raised in primary and secondary school contexts, particularly with regard to the need for further teacher training to ensure that curricula change is meaningful.

ESE—only for outdoor learning?

There has always been a very strong relationship made between ESE and the natural world, positioning nature-based learning as a recognisable feature of ESE (Inoue, 2015; Lee & Kim, 2017; White et al., 2018; Wilson, 2012; Young & Elliott, 2003). It is argued that historically, ESE has been narrowly grouped alongside outdoor programmes, outdoor research and excursions (Lee & Kim, 2017) rather than more formal, classroom-based learning. Furthermore, there is the tendency for school-based ESE programmes, which are often conducted outdoors, to miss opportunities for learners to use problem and inquiry-based learning to make meaning from what they are learning. White et al. (2018) contend:

> To date, outdoor school-based environmental education has typically focused on school ground improvement and greening initiatives, horticultural projects, and outdoor play developments, but have only recently become the focus of empirical inquiry.
>
> *(p. 3)*

In response, they developed and evaluated an urban school-based ESE project to increase learners' awareness and knowledge about the environment, given their argument that many children in urban areas are disconnected from the environment. Across eight primary schools in the UK, consisting of 220 children participants aged 7–10, they found that the six-week programme, *Bird Buddies*, increased ecological awareness and bird identification knowledge. While classed as an 'intervention', they argued that after the project, several schools were continuing to watch and feed birds. However, the project highlights that local, long-term approaches are needed to make ESE a key feature of learner learning.

However, Chan, Mathews and Li (2018) argue that the success of a long-term ESE initiative, focused on the Caohai Nature Reserve in southwest China, was due to the strong relationships and social capital built among local schools, nature reserve staff, farmers, non-profit organisations and government entities. While Chan et al. (2018) argue that integrating ESE into schools can be difficult in China, given its exam-oriented schools, it took the cooperation and trust of several different groups of people to create a united front that would educate both learners and teachers on the degradation of the biodiversity and agriculture of the Caohai Nature Reserve in order to make learning locally meaningful.

ESE curriculum approaches—infusion, cross-curricular themes and subject-specific

While Ramsey, Hungerford and Volk (1992) emphasised the need for ESE to be infused into the K-12 curriculum several decades ago, positioning ESE into the curriculum has taken on many forms. Infusion, or 'the integration of content and skills into existing courses in a manner as to focus on the content (and/or skills) without jeopardizing the integrity of the courses themselves' (Ramsey et al., 1992, p. 40), in many ways has evolved into this notion of a cross-curricular approach. Parker (2017) reports on how Indonesia's new curriculum embeds ESE into religious education. She argues that in Indonesia, given that the environment is embedded into religious concepts such as creationism (God created the world) and instrumentalism (the environment was made for humans), there were missed opportunities to highlight human responsibility or agency in protecting and caring for the earth. Parker (2017) argues that while framing the environment with religious ideals can be effective, there needs to be a stronger focus on 'human responsibility for the destruction of the natural world' (p. 1267). As suggested above, ESE themes and concepts have been embedded into single subjects and even religious ideals, but there has also been increasing attention on ESE as a cross-curricular theme that connects across several subject areas.

Tedesco, Opertti, and Amadio (2014) found that at least 70 countries included cross-cutting or transversal themes in their curricula as a way to link content across a number of disciplines. Australia, for example, introduced three Cross-Curricular Priorities (CCPs) in the Australian (national) curriculum, with one of them being sustainability (ACARA, 2020a). While the development of the curriculum began in 2009, it was implemented in all Australian states and territories in 2014 (ACARA, 2020b). However, the notion of CCPs occupies a 'perennially precarious space' (Salter & Maxwell, 2015, p. 2), particularly because they are positioned as a priority but spread across a number of learning areas with limited accountability and/or explicit requirements for how or to what extent it is taught. Kuzich, Taylor and Taylor (2015) found that teachers often ignore ESE in the curriculum in Australia because it is difficult to identify ESE elements and then know how to integrate them into the current content. While sustainability is flagged in the curriculum document by a leaf icon, there is limited direction and support to aid teachers who may lack the content knowledge in ESE to successfully implement this in their teaching without a lot of additional time and effort.

In Korea, ESE has been positioned in the primary, middle and high school curricula since 1981, but it evolved from being a CCP to becoming its own subject, 'Environment,' in 1992—which was unprecedented at that time (Lee & Kim, 2017). However, while Korea has led the world in promoting ESE in their national curriculum since 1981, Lee and Kim (2017) argue that even though 'Environment' is now offered as a subject in the national curriculum, not all schools have selected to offer 'Environment' as a subject option. Countries like India and Brazil have a 'right to the environment' and 'protection of the environment' as both a right and a duty embedded within their constitutions. However, similar to Korea, the embedding of ESE within the curriculum in India particularly, the EC curriculum has been limited despite the huge potential of reaching out to the largest population of young children in the world (Almeida & Ohara, 2020).

While a number of curriculum initiatives in primary and secondary schools have either placed ESE as a theme or concept within another learning area, subject or religion (e.g. Parker, 2017), as a CCP (e.g. ACARA, 2020a; Barnes, Moore & Almeida, 2019; Lee & Kim, 2017) or a stand-alone subject (e.g. Lee & Kim, 2017), there are still challenges and concerns that inhibit ESE being implemented in schools and allowing for meaningful change.

What next? How to address current challenges and concerns in implementing ESE

A number of studies have explored secondary learners' views on sustainability and/or environmental issues (e.g. Jackson et al., 2016; Ntona, Arabatzis & Kyriakopoulos, 2015) as well as primary and secondary teachers' implementation of ESE (e.g. Barnes et al., 2019; Yeung, Lee & Lam, 2012); however, many of these studies highlight that there are significant policy-practice gaps that make meaningful and long-term changes difficult. While several studies report that teachers generally hold positive views towards ESE, there are concerns related to how to (1) successfully tackle patterns of learner behaviour towards sustainability (Jackson et al., 2016; Ntona et al., 2015) and (2) provide the structure and support for ESE implementation within a crowded curriculum, full workload and/or testing culture (Barnes et al., 2019; Chan et al., 2018; Yeung et al., 2012).

Jackson et al. (2016) quantitatively explored the environmental attitudes of 483 secondary learners studying in two international and two local schools in Hong Kong. While they found that there was no significant difference in attitudes or behaviours between learners in the international schools as opposed to local schools, they did find that learners' environmental knowledge, attitudes and behaviours are significantly correlated with age and gender. In other words, similar attitudes and behaviours are held by those within the same age group and/or gender. This suggests that more qualitative research is needed on the environmental knowledge, attitudes and behaviours of particular age groups so that curricular initiatives can be targeted and be more meaningful and relevant to learners.

In addition, to responding to learners' attitudes and behaviours as ways to address and make ESE contextually relevant and meaningful, a number of studies have raised concerns that for ESE to be effectively implemented in schools globally, there needs to be a whole-school approach that equips and scaffolds teachers with the time and resources required (Barnes et al., 2019; Chan et al., 2018; Lee & Kim, 2017), high-quality teacher training (Chan et al., 2018; Lee & Kim, 2017) and a project-based and/or inquiry-based approach (Littledyke et al., 2009) in light of the limited structure and support that is currently available in schools.

The next section will explore some of these concerns within the context of Australia, particularly as national initiatives have galvanised the position of ESE in Australian schools; yet its impact within schools has been seemingly limited.

ESE implementation in Australia's educational settings

Over the last several decades, Australia has promoted a number of ESE initiatives which have positioned ESE as an important curricular objective and/or goal within Australian educational settings. In light of these initiatives, this section will describe one study, conducted within an Australian state/territory, to highlight how, despite a number of positive ESE policy initiatives, they have not always resulted in meaningful curriculum change.

A time for change

Acknowledging the increasing global concern regarding the conservation of the environment (Gough, 2011), the Australian government responded with numerous policy documents, reports and strategies to address the need to create a sustainable world (Davis, 2010). Influenced by global trends in establishing sustainability policies, Australia introduced the *National Conservation Strategy for Australia* in 1984, which focused on educating communities towards sustainable development and conservation (Gough, 2011). In 2000, the government established the *National Advisory Council* which recognised the urgent need for ESE and provided a national action plan that would lay the foundations for participation in the UN *Decade of Education for Sustainable Development*, 2005–2014, and trigger 'agenda setting and concrete actions around sustainability and EfS' (Davis, 2010, p. 11). The National Review of Environmental Education and its contribution to sustainability in Australia provided a comprehensive report of Environmental Initiatives across various sectors in Australia. Its review of the school sector showed limited uptake of policies and strongly recommended whole-school approaches (ARIES, 2005).

Following this review, the national policy document, *Melbourne Declaration on Educational Goals for Young Australians* (2008), positioned sustainability as an education goal. Additionally, in response to the global financial crisis of 2007–2008, the Australian government established a number of initiatives aimed at investing in important infrastructure for education (Wettenhall, 2011). While there were several building funds offered for Australian schools, the government also announced the National Solar Schools Program (NSSP), which offered primary and secondary schools grants to install solar and other renewable systems. One of the conditions of this grant was to ensure that the use of these tangible resources also lead to educational impact. More specifically, one of the five objectives of NSSP was to 'allow schools to provide educational benefits for school learners and their communities' (Grosvenor Management Consulting, 2013, p. 85). As a result, Australian states and territories had to choose a preferred Data Collection, Storage, Visualisations System (DCSVS) that would be employed in their local schools and early childhood settings (Grosvenor Management Consulting, 2013) that could be used as an educational resource. The DCSVS was to record, store and display data for electricity generated from solar panels and the centre or school's consumption of water and gas.

With the NSSP and the *Melbourne Declaration on Educational Goals for Young Australians* (2008) laying the foundation for an ESE focus in Australian schools, sustainability became a 'cross-curricular priority' within the Australian Curriculum—Australia's first national curriculum (ACARA, 2020a). Sharing cross-curricular status with two other content areas—Aboriginal and Torres Strait Islander histories and culture and Asia and Australia's engagement with Asia—the purpose of having CCPs in the curriculum was to place these three content areas across several learning areas. Salter and Maxwell (2015) suggest:

> ...these priorities are uniquely positioned in the emerging curriculum as contexts for learning supported by organising ideas that establish knowledge, understandings and skills for each priority... Despite, or perhaps due to, the uniqueness of the CCPs they occupy a perennially precarious space in the emerging Australian Curriculum.
>
> *(p. 2)*

Therefore, positioning the CCPs as a 'priority' but spreading them across a number of learning areas results in limited explicit instructions or requirements for how these priorities are to be taught. Salter and Maxwell (2015) further explain:

> The construction of the priorities as (optional) solutions to problems, rather than intrinsically worthwhile content, has meant that bodies such as ACARA could respond to stakeholder demands that such content be included or excluded as politically expedient. Rather than being developed

with the interests of school communities in mind, the priorities have been engineered to address the interests of others.

(p. 14)

Consequently, even with the educative push for sustainability as a CCP, a national report on the impact of NSSP in 2013, prepared by the Department of Resources Energy Tourism (DRET) and Grosvenor Management Consulting, found that less than 50% of the schools surveyed across Australia were incorporating the subject of energy efficiency in their learning materials (2013, p. 88). However, with a glimpse of hope, the report also reported that 'while only 35% of survey respondents utilized the DCSVS data in their lessons plans, almost 50% of schools were planning to do so in the near future' (Grosvenor Management Consulting, 2013, p. 91).

The following section revisits the impact of the NSSP in promoting ESE in Australian classrooms through the evaluation of one specific Australian State/Territory Government's DCSVS. The study highlights that while building infrastructure and resources is important in addressing ESE in education, there is still the need to explore steps forward to promote sustainability in schools/centres.

A look at one ESE initiative in one state/territory

In 2016, the authors conducted a small-scale study to evaluate how one DCSVS was promoted and implemented by teachers and school leaders. The study aimed to capture the perspectives of early years, primary and secondary teachers and school leaders (e.g. principals, subject leaders, business managers) with regard to the implementation of the DCSVS through an online survey, and two surveys were created using Qualtrics, an online survey platform, and utilised both Likert-scale questions and open-ended questions. A survey was created for teachers and another for school leaders, and they were distributed to 86 local schools and centres in July 2016. The survey captured the perspectives of 116 respondents, 66 teachers and 50 school leaders.

The study's most revealing finding was that 82% (n = 54) of the teacher participants were not aware of the website, with 9% (n = 6) knowing about the website but not currently using it and only 9% (n = 6) using the website and finding it useful. While at first the findings were disheartening, the authors argue that (1) ESE implementation begins with leadership and (2) there is still hope.

Many of the school leaders in this study found that the DCSVS was useful but viewed its value with regard to resource management rather than for educational benefits. For example, when the school leaders were asked to select the purposes of the DCSVS that were applicable to their school, 66% (n = 33) reported they used the system to identify problems (e.g. water leaks) and 32% (n = 16) to improve energy efficiency and reduce energy consumption. However, 18% (n = 9) claimed that they did not use the website for any particular purpose and 22% (n = 11) did not know the website existed. Additionally, many school leaders acknowledged that the system has value but was simply not a high priority for their schools:

> In all honesty professional learning for staff has been mainly around Literacy, Numeracy and Special Needs education specifically.

> We focus on literacy and numeracy along with learner welfare and engaging parents. I'm not saying there is anything wrong with the website but it is a low priority…No offence but this website is not core business so I'll avoid it as long as I can.

Within an education landscape where teachers and school leaders are feeling the strain of an overcrowded curriculum (Caldwell, 2015) and/or the inability to find the time needed to explore new approaches and resources to support learning, the findings from this study suggest that we need to consider a multifaceted, multi-modal approach to the learning and teaching of ESE. With the establishment of federally funded DCSVS programmes in Australia, there are numerous resources available, but these resources need to be coupled with explicit, targeted and practical support for ease and meaningful implementation at both the whole-school level as well as the classroom level.

Without losing hope, the study also found that a number of teachers, particularly those who were unaware of the DCSVS before the survey, were optimistic about using the resource in their classrooms. This highlights that while the DCSVS initiative may not have resulted in substantive change as measured by a whole-school approach to ESE, change begins with one person and that creating a buzz and/or interest is vital in creating change. Rowe and Liemer (2002) argue that one small step can be the catalyst for substantive change, mirroring the idea that the moving of one small stone can result in a landslide. Therefore, while our previous findings suggest that curricular change is slow, complex and at times appears to be thwarted, we choose to frame this book with the hope that small changes, steps, initiatives and discussions are all important aspects of promoting change.

References

ACARA. (2020a). Sustainability. Retrieved from https://www.australiancurriculum. edu.au/f-10-curriculum/cross-curriculum-priorities/sustainability/

ACARA. (2020b). About the Australian curriculum. Retrieved from https://www. australiancurriculum.edu.au/about-the-australian-curriculu

Almeida, S. C. (2015). *Environmental education in a climate of reform*. Rotterdam, The Netherlands: Sense Publishers.

Almeida, S. C., & Ohara, Y. (2020). Research in early childhood education for sustainability: Policies and perspectives from India. In S. Elliott, E. Ärlemalm-Hagsér & J. Davis (Eds.), *Researching early childhood education for sustainability: Challenges, assumptions and orthodoxies* (pp. 82–93). New York, NY: Routledge.

Ärlemalm-Hagsér, E., & Engdahl, I. (2015). Careing for oneself, other and the environment: Education for sustainability in Swedish preschools. In J. Davis (Ed.), *Young children and the environment: Early education for sustainability* (pp. 251–262). Port Melbourne, Australia: Cambridge University Press.

Australian Research Institute for Environment and Sustainability (ARIES). (2005). *A national review of environmental education and its contribution to sustainability in Australia.* Retrieved from http://aries.mq.edu.au/projects/national_review

Barnes, M., Moore, D., & Almeida, S. (2019). Sustainability in Australian schools: A cross-curriculum priority? *Prospects, 47*(4), 377–392. doi:10.1007/s11125-018-9437-x

Barth, M., & Rieckmann, M. (2012). Academic staff development as a catalyst for curriculum change towards education for sustainable development: An output perspective. *Journal of Cleaner Production, 26*, 28–36.

Barth, M., & Rieckmann, M. (2016). State of the art in research on higher education for sustainable development. In M. Barth, G. Michelsen, M. Rieckmann, & I. Thomas (Eds.), *Routledge handbook of higher education for sustainable development* (pp. 100–113). London, UK: Routledge.

Caldwell, B. (2015). Feeling overwhelmed? It is time for serious innovation. *Australian Educational Leader, 37*(1), 14–17.

Chan, Y-W., Mathews, N., & Li, F. (2018). Environmental education in nature reserve areas in southwestern China: What do we learn from Caohai? *Applied Environmental Education & Communication, 17*(2), 174–185. doi:10.1080/1533015X.2017.1388198

Davis, J. (2010). Chapter 1. What is early childhood education for sustainability. In J. Davis, (Ed.), *Young children and the environment: Early education for sustainability* (pp. —21–42). Cambridge: Cambridge University Press.

Davis, J. (2015). What is early childhood education for sustainability and why does it matter? In J. Davis (Ed.), *Young children and the environment: Early education for sustainability* (pp. 7–27). Port Melbourne, Australia: Cambridge University Press.

Elliott, S. (2014). *Sustainability and the early years learning framework.* Mt Victoria, Australia: Pademelon Press.

Elliot, S., Ärlemalm-Hagsér, E., & Davis, J. (2020). *Researching early childhood education for sustainability: Challenging assumptions and orthodoxies.* New York, NY: Routledge.

Elliott, S., & Davis, J. (2018). Chapter 12. Moving forward from the margins: Education for sustainability in Australian early childhood contexts. In Reis, G. & Scott, J. (Eds.), *International perspectives on the theory and practice of environmental education: A reader (Environmental Discourses in Science Education, Volume 3)* (pp. 163–178). Switzerland: Springer.

Gough, A. (2011) The Australian-ness of curriculum jigsaws: Where does environmental education fit? *Australian Journal of Education, 27*(1), 9–23.

Grosvenor Management Consulting. (2013). *National solar schools program evaluation report.* Retrieved on November 14, 2016 from http://www.industry.gov.au/Energy/EnergyEfficiency/Documents/NSSP-Evaluation- Report-Final.pdf

Inoue, M. (2015). Beyond traditional nature-based activities to education for sustainability: A case study from Japan. In J. Davis (Ed.), *Young children and the environment: Early education for sustainability* (pp. 264–274). Port Melbourne, Australia: Cambridge University Press.

Jackson, L., Pang, M. F., Brown, E., Cain, S., Dingle, C., & Bonebrake, T. (2016). Environmental attitudes and behaviours among secondary learners in Hong Kong. *International Journal of Comparative Education and Development, 18*(2), 70–80.

Ji, O. (2015). Education for sustainable development in early childhood in Korea. In J. Davis (Ed.), *Young children and the environment: Early education for sustainability* (pp. 264–274). Port Melbourne, Australia: Cambridge University Press.

Karim, S., Srisandarajah, N., & Heiter, A. (2013). Systems study of an international master's program: A case from Sweden. In L. Johnston (Ed.), *Higher education for sustainability: Cases, challenges, and opportunities from across the curriculum* (pp. 63–78). Retrieved from https://ebookcentral-proquest-com.ezproxy.lib.monash.edu.au

Kuzich, S., Taylor, E., & Taylor, P. C. (2015). When policy and infrastructure provisions are exemplary but still insufficient. *Journal of Education for Sustainable Development, 9*(2), 179–195.

Lee, S-K., & Kim, N. (2017). Environmental education in schools of Korea: Context, development and challenges. *Japanese Journal of Environmental Education, 26*(4), 7–14.

Li, J. (2013). Environmental education in China's college English context: A pilot study. *International Research in Geographical and Environmental Education, 22*(2), 139–154. doi: 10.1080/10382046.2013.779124

Littledyke, M., Taylor, N., & Eames, C. (2009). *Education for sustainability in the primary classroom.* South Yarra, VIC: Palgrave MacMillan.

Livingstone, R. (2016, August 8). ACECQA *We hear you blog: Demystifying sustainability.* Retrieved from https://wehearyou.acecqa.gov.au/2016/08/08/demystifying-sustainability/

Ministerial Council on Education, Employment, Training and Youth Affairs. (2008). *Melbourne declaration on educational goals for young Australians.* Retrieved from http://www.curriculum.edu.au/verve/_resources/national_declaration_on_the_educational_goals_for_young_australians.pdf

Munoz-Pedreros, A. (2014). Environmental education in Chile: A pending task. *Ambiente & Sociedade, 17*(3), 175–194.

National Solar Schools Program (NSSP). (2008). *Solar for schools.* Retrieved from http://www.energymatters.com.au/solar-for-schools/school-solar-grant/

Ntona, E., Arabatzis, G., & Kyriakopoulos, G. (2015). Energy saving: Views and attitudes of learners in secondary education. *Renewable and Sustainable Energy Reviews, 46*, 1–15. doi:10.1016/j.rser.2015.02.033

Parker, L. (2017) Religious environmental education? The new school curriculum in Indonesia. *Environmental Education Research, 23*(9), 1249–1272. doi:10.1080/13504622.2016.1150425

Powers, A. (2004). Teacher preparation for environmental education: Faculty perspectives on the infusion of environmental education into preservice methods courses. *Journal of Environmental Education, 35*(3), 3–11.

Ramsey, J., Hungerford, H., & Volk, T. (1992). Environmental education in the K-12 curriculum: Finding a niche. *Journal of Environmental Education, 23*(2), 35–45. doi:10.1080/00958964.1992.9942794

Rowe, D., & Johnston, L. (2013). Learning outcomes: An international comparison of countries and declarations. In L. Johnston (Ed.), *Higher education for sustainability: Cases, challenges, and opportunities from across the curriculum* (pp. 45–60). Retrieved from https://ebookcentral-proquest-com.ezproxy.lib.monash.edu.au

Rowe, S., & Liemer, S. (2002). One small step: Beginning the process of institutional change to integrate the law school curriculum. *Journal of the Association of Legal Writing Directors, 1*, 218–229. Retrieved on SSRN from https://ssrn.com/abstract=1095632

Salter, P., & Maxwell, J. (2015). The inherent vulnerability of the Australian curriculum's cross-curriculum priorities. *Critical Studies in Education, 56*(2), 1–17.

Sprenger, S., & Nienaber, B. (2018). (Education for) sustainable development in geography education: Review and outlook from a perspective of Germany. *Journal of Geography in Higher Education, 42*(2), 157–173. doi:10.1080/03098265.2017.1379057

Tedesco, J. C., Opertti, R., & Amadio, M. (2014). The curriculum debate: Why it is important today. *Prospects, 44*, 527–546.

Tejedor, G., Segalàs, J., & Rosas-Casals, M. (2018). Transdisciplinarity in higher education for sustainability: How discourses are approached in engineering education. *Journal of Cleaner Production, 175*, 29–37. doi:10.1016/j.jclepro.2017.11.085

Vincent, S., & Focht, W. (2011). Interdisciplinary environmental education: Elements of field identity and curriculum design. *Journal of Environmental Studies and Science, 1*, 14–35. doi:10.1007/s13412-011-0007-2

Weldemariam, K., & Wals, A. (2020). From an autonomous child to an entangled child within an agentic world: Implication for early childhood education for sustainability. In S. Elliot, E. Ärlemalm-Hagsér, & J. Davis (Eds.), *Researching early childhood education for sustainability: Challenging assumptions and orthodoxies* (pp. 13–24). New York, NY: Routledge.

Wettenhall, R. (2011). Global financial crisis: The Australian experience in international perspective. *Public Organization Review, 11*, 77–91. doi:10.1007/s11115-010-0149-9

White, R., Eberstein, K., & Scott, D. (2018). Birds in the playground: Evaluating the effectiveness of an urban environmental education project in enhancing children's awareness, knowledge and attitudes towards local wildlife. *PLoS One, 13*(3), 1–23. doi:10.1371/journal.pone.0193993

Wilson, R. (2012). *Nature and young children* (2nd ed.). London, UK: Routledge.

Yeung, Y-Y., Lee, Y-C., & Lam, I. (2012). Curriculum reform and restructuring of senior secondary science education in Hong Kong: Teachers' perception and implications. *Asia-Pacific Forum on Science Learning and Teaching, 13*(2), 1–34.

Yi Lo, E. Y. (2010). Environmental education in Hong Kong kindergartens: What happened to the blue sky? *Early Child Development and Care, 180*(5), 571–583. doi:10.1080/03004430802181361

Young, T., & Elliott, S. (2003). *Just discover! Connecting young children with the natural world.* Croydon, VIC: Tertiary Press.

3

IDENTITIES MATTER

Teachers' identities as a lens into teachers' everyday practices

This study aims to understand how early career teachers, who are new to their teaching careers, practice and integrate Environmental and Sustainability Education (ESE) into their everyday teaching practices. Who we are is about what and how we teach. In order to fulfil this aim, we need to look at individuals and how their individual identities support their practices. Identities are shared sets of meanings that define individuals and have become fundamental aspects when researching teaching (Akkermann & Meijer, 2011). These are attained over time and in specific socio-cultural political contexts and are basically a 'way of defining, describing and locating oneself' (Clayton, 2012, p. 2).

Identity defines the most vital attributes of human distinctiveness through assimilating the simultaneous experiencing of self-sameness on the one hand and self-distinction on the other (Erikson, 1956). The origin of the word 'Identity' is derived from two Latin terms: *Idem* and *Ipse* (Ricoeur, 1992, p. 2). Ricoeur (1992, pp. 2–3) argues that the concept of 'identity,' in fact, has more complex origins. According to him 'identity,' when observed from its Latin origin term *Ipse*, provides a relatively separate view of the word 'sameness.' He considers it the sense of selfhood (*ipséité*) – meaning the more intimate, 'who the Self is' type of self- identity (Meltzer, 2010, p. 517; Ricoeur, 1988, p. 246). When seen from its other Latin origin term *Idem*, the theoretical understanding of the concept 'identity' alludes to 'essence,' 'being' and 'change.' The second (*ipseite*) descent of the phrase implies the individual in contrast—oneself as resisted to the other(s): Ego-Alter Ego, Us-Them, Oneself-Another and even in contrast to the individual, that is, 'Oneself as Another' (Dauenhauer & Pellauer, 2014; Ricoeur, 1992).

James (1892/1961, p. 44 as cited in Clayton & Myers, 2015, p. 2), described a tripartite categorisation between the material, social and personal aspects of the 'objective' self, or 'the sum total of all a person can call his' which included the concept of the dynamic, instinctive part of identity. The core interest within this stream of philosophical thought is the building and sustaining of human identity in the context of society.

Identity development is a complex concept, particularly depending on one's circumstances, knowledge, maturity and other factors. They evolve over time based on the context of every individual's socio-cultural and political realities (Clayton, 2012; Ryan & Deci, 2003). These are not fixed in time and place, but as something in the process of becoming rather than being, although 'the past continues to speak to us' (Hall, 1990, p. 395). Erikson developed a comprehensive identity theory that acknowledged identity formation as a result of interaction between psychic and social factors, which were always embedded in a larger cultural context and give individuals a sense of who they are (Erikson, 1968). Identity grounds us and gives us meaning (Spencer & Markstrom-Adams, 1990). Yet, our identity is an ever-changing process, continuously changing (Hall, 1996), based on internal constructions (and reconstructions) not an 'outside discourse' (Hokowhitu, 2010, p. 4). Elsewhere, identity is also described as, 'points of identification and attachment, only because of their capacity to exclude, to leave out, to render "outside" [the] abjected' (Hokowhitu, 2010, p. 5). Still, for us to function at our best, our understanding of our identity needs to be clear in our minds and include our goals, values and beliefs (Waterman, 1985).

According to Erikson (1968), identity consists of components like self-esteem, self-development, self-perception, self-recognition, self-awareness and a locus of control. Social influence enhances and sharpens identities, so social implications of identities and relations with others become more important (Clayton, 2012, p. 5). Erikson's timeless quote describes the important dynamics between the internal psychological, developmental and social 'sharing' nature of identity:

> The term 'identity' expresses such a mutual relation in that it connotes both a persistent sameness within oneself (self-sameness) and a persistent sharing of some kind of essential characteristics with others.
>
> *(Erikson, 1956, p. 57)*

Individuals have multiple identities based on their roles in society (role identity), specific groups (group identity) and as individuals (personal identities) (Stets & Serpe, 2013). Identity type and contextual social settings impact an individual (Brenner, Serpe & Stryker, 2014).

Teacher identities are complex and shaped by the various interpretations of their 'personalities, values, actions, and sense of self' present to their social relationships (Thomashow, 1996). Teacher identity development therefore is a continual, dynamic and reinventing process, continually adjusting, based on internal and external factors (Schutz, Nichols & Schwenke, 2018). Understanding their identities is, therefore, crucial in understanding their teaching practices. A focus on identities acknowledges that teachers' work is complex, emotional, requiring courage and passion and is more than just acquiring assets like knowledge, competencies and beliefs (Akkerman & Meijer, 2011). Emphasis on identity also allows us to acknowledge the teacher as the centre—the starting point (Akkerman & Meijer, 2011)—thereby providing a more holistic perspective of what matters in ESE implementation. It also aligns with Sachs (2005) positioning of teacher identity as the core of the teaching profession, providing a structured framework for teachers to develop their own ideas of how they define their work.

Three particular aspects of their identities, namely their personal identities, their professional identities as teachers and their environmental identities, stand out in this study. They determine how these teachers negotiate and implement ESE in their current roles as early career teachers in their first teaching positions.

Analysing their ESE practices using these three lenses shed light on early career teachers' experiences in their educational settings and their shifting identities. They help to understand aspects of their teaching that work towards promoting ESE, while at the same time highlighting issues and constraints that limit its implementation. This centres teachers identities in line with the recommendations from a recent study that found teachers' sense of agency that is central to their identities as ESE actors is often threatened, and therefore it is necessary to keep this as the focus of concern while formulating new policies (Zaradez, Sela-Sheffy & Tal, 2020). This chapter engages with identities as a framework to understand how these early career teachers made sense of who they were and how that shaped their teaching practices.

In this study, our participants were selected based on strong environmental values in their personal identities. This was witnessed by the authors who worked with these, then pre-service, teachers over the course of a few years. They were passionate pre-service teachers who worked closely with their educators and showed a strong affinity for environmental learning, justice and action. It was clear from their engagements during their pre-service learning times that they had intentions to foreground ESE when they started their position as teachers. The main aim of identity theory is to specify how the meanings attached to various identities are negotiated and managed in interaction (Stets & Serpe, 2013, p. 31). Teachers' thinking and deeply held beliefs are the foundations on which teachers' action and implementation of ESE are based (Hart, 2003). This aligns with our study goals to understand how strong environmental identities are invoked, negotiated and managed as these new graduate teachers embarked on their professional teaching careers. According to Milstein (2020), every identity is ecological, and it is crucial we see the interlinkages and impacts of these connections in relation to who we are as individuals. In this chapter, we discuss teachers' personal and professional identities and the interplay with their environmental identities.

Personal identities

An individual's personal identity can be understood as the 'meanings that individuals hold for themselves – what it means to be who they are' (Burke, 2003, p. 2). The self is made of many identities (Stets & Burke, 2014). In other words, individuals have multiple identities that influence behaviour, with some identities to be invoked more often than others (Jones & McEwen, 2000; Werbner, 2017). Stryker (1980) asserts that some aspects of an individual's identity are more prominent and salient when compared with others. These conceptions are central to understanding how these early career teacher participants perceived their identities, as it then allows for an analysis of key influences that shape their identities and selves. Personal identities are unique to an individual, stay with a person always and set them apart from others in a group. Due to their high salience and commitment, these are also viewed as 'master identities' that influence all other identities (Stets & Burke, 2014).

Identity is seen as something that develops in social contexts, is fluid or everchanging and is something that one develops over the course of one's life. In other words, as Beijaard, Meijer and Verloop (2004) summarise, 'Identity development occurs in an intersubjective field … is an ongoing process of interpreting oneself as a certain kind of person … in a given context. Identity answers the recurrent question "Who am I at this moment?"' (p. 108). Identity is varied and multiple, it is always being made and shifts based on contexts and relationships (Rodgers & Scott, 2008).

Rodgers and Scott (2008) position identity as contextual, relational and emotional as well as storied with an inherent awareness and voice. One way to make sense of these multifaceted identities is through personal narratives which offer opportunities to tell stories. Connelly and Clandinin (1999) argue that teachers' identities are best understood through stories that offer modes of embodiment of their lived experiences and the scenarios that shaped their past, their present and their future. Teachers' self, as the meaning maker and their identity as the meaning made (Rodgers & Scott, 2008), are sought to be captured through their stories. This supports our use of narrative inquiry as a methodology allowing for our early career teachers' stories and often marginalised voices to be heard. Personal identities are building blocks for other identities that are developed through every individual's lived experiences.

Professional identities

Professional identities emerge in the context of one's professional roles. Professional identities connect to ways in which individuals become teachers, how they understand the profession and how their teaching is reshaped with their everyday experiences (Almeida, 2015). An individual's personal or 'master identity' influence all spheres of their lives (Stets & Burke, 2014), including their professional roles as teacher. Identity theory contends that identities are intimately tied to social structure. Individual actions are geared towards verifying their identities while maintaining the social system. This verification is done through the allocation of actual (available) and potential (future) resources, and every effort is made to define and maintain the social system in the face of distractions and disruptions (Stets & Burke, 2014). An individual's personal identity supports them in making these adjustments to their professional selves, allowing them to function in the new social system of a school or an early childhood centre. Day (2018, p. 68) contends that 'to succeed over time as professionals, teachers need to have and sustain a positive sense of professional identity.' This is especially true for early career teachers who are new to these settings and are constantly making adjustments to fit into the existing structures.

Beijaard et al. (2004, p. 122) identified four features that are essential to professional identity formation:

1. Professional identity formation is an ongoing process of interpretation and reinterpretation of experience
2. Professional identity implies both person and context
3. A teacher's professional identity consists of multiple sub-identities (or 'selves') that more or less harmonise. The more central a sub-identity is, the more costly it is to change or lose that identity
4. Agency is an important element of professional identity, meaning that teachers have to be active in their processes of professional learning.

Early career teachers' professional identities are, therefore, being constantly moulded by their experiences in educational contexts, with a constant reinterpretation of their personal identities, a search for their voice and agency as they engage with professional learning processes. Given that they are new to the system, these identities rely heavily on context. Educational settings, policy structures, organisational support and mentor teachers all play a crucial role as these early career teachers actively engage with an apprenticeship of observation (Lortie, 1975). Previous research (Almeida, Moore & Barnes, 2018) has called for a focus on building capacity and individual strengths rather than mere resources and deep learning professional development opportunities that have lasting impact on teachers' professional identities.

Ecological/environmental identities

The above discussion on professional and personal identities foregrounds the role of 'sub-identities' and the need to harmonise these. This synchronisation is especially necessary for central identities—changing or losing these is harder and more disruptive. In this study, all our participants had been identified as having strong environmental identities based on their earlier interactions with the authors as pre-service teachers. We did not use any scale to measure this but relied on our conversations and interactions with these participants to recognise this during our time of teaching them and also during the first introductory workshop held for this research project. There are a range of terms—from environmental identities to ecological identities to eco-identities—used variably to depict the connections between the environment and our identities.

Environmental identities are about how individuals orient themselves to the natural world and determine an individual's action based on a sense of who they are (see Clayton and Opotow, 2003). Clayton (2012) provides an early definition of environmental identity as:

> a sense of connection to some part of the nonhuman natural environment, based on history, emotional attachment, and/or similarity, that affects the ways in which we perceive and act toward the world; a belief that the environment is important to us and an important part of who we are.
>
> *(pp. 45–46)*

Thomashow (1996, p. 3) postulates ecological identities as myriad ways in which individuals see themselves 'in relationship to the earth as manifested in personality, values, actions, and sense of self.' A strong environmental identity can transcend personal, professional, social and political aspects of an individual. It is built on social interactions heavily influenced by and contributing to social and cultural meaning. As with all other identities, these are developed and evolved over an individual's lifetime based on significant life experiences, interactions with others and socially constructed understandings. Environmental identities are strongly linked to a person's connection with the earth, perception of the ecosystem and direct experience of nature. Social identities, in particular gender and race/ethnicity, also have a significant impact on an individual's environmental identity development (Miao & Cagle, 2020).

However, it is important to deviate a little from the known definitions of environmental identities and eco-identities. These identities are seen to emerge through strong non-human nature connections where nature might be seen as green, pristine and something 'out there.' Existing understandings of environmental identities exclude people who have not grown up in nature. Personal experiences of growing up in a heavily urbanised, crowded city like Mumbai have reshaped understandings of what the term 'nature' means. This is supported by research where 'a garden on a window sill' may be the only 'nature' experiences available to individuals in their everyday lives (Sageidet, Almeida & Dunkley, 2018). Nature could be simply that of one mango tree in the school yard that one rushes to climb during every lunch break to find peace and a sense of calm in a highly crowded environment. Nature could also be the under-construction buildings full of concrete, bricks and mud that were the only available playgrounds for children in the locality. These different identities need to be acknowledged. It needs to be legitimated that pristine 'nature' experiences are not a precursor; that individuals growing up in heavily urban environments can still develop strong environmental identities and connections to the place. Teachers in our project did not discuss these different nature experiences, but it is important to acknowledge these differences as we conceptualise understandings of 'environmental identities' (Almeida, 2020).

Environmental identities can help explain the drive that motivates people towards uptake of environmentalism in their personal and professional lives (Almeida, 2015). For teachers, this translates into implementation of ESE in their teaching practices.

The confluence—where three identities meet

To summarise the earlier three sections, an identity is the way an individual views the self. Identities are multiple, complex, multi-layered and vary in salience as well as importance, depending on the immediate context and past experiences. They are experienced internally as well as externally, and impact how we respond to the world, cognitively and emotionally (Clayton & Myers, 2015). While there are many facets or multiple identities that any individual carries, for the purposes of this study we focus on teachers' personal, professional and environmental identities, as this provides a lens to help answer our research questions, as articulated in Chapter 5. Teachers' identity development processes often reflect struggles between their diverse identities and a constant adjustment/balance for coherence and continuity. In Akkerman and Meijer's (2011) words, it reflects the struggles of being one and being many at the same time (p. 318). Discussions that support asking questions about 'how to be someone who teaches, not just about how to teach' are crucial in helping support teacher identity development (Eliasson, 2019). A way forward is to examine the ways in which teachers work with and around their struggles and uncertainties in their daily practices and how their multiple identities merge. This allows us to have a deeper insight into how teachers' identities are developed and supported in their early years of teaching.

We conceptualise the coming together of these three identities as a confluence or 'sangam' (in Indian languages) of three rivers, as seen in the image below

FIGURE 3.1 The confluence of rivers (Stan Dalone, 2009) 'File:Sotočje Vipave in Hublja. Confluence of rivers Vipava and Hubelj. (3282732600).jpg'

(Figure 3.1). The term 'confluence' is defined as an 'act or process of merging' and 'a coming or flowing together, meeting, or gathering at one point' (https://www.merriam-webster.com/dictionary/confluence). In India, this point of merger or 'sangam' is considered special and sacred. Sangam is where some of India's holiest temples stand and millions of people converge here for the holy bath that provides hope of a better afterlife.

Figure 3.2 shows the process of the merging of these multiple identities allowing for a sangam. This is the special way in which identities are represented through/across/between blurred boundaries where ideas flow and intermingle. Here identities meet to further strengthen the flow and strength of the individual identities. Teachers with strong personal experiences with the natural environment and environmental problems—either positive, place-based experiences or negative experiences of environmental destruction—go on to develop strong environmental identities which impact their response to environmental information and issues. This response is then displayed through their professional identities, as it informs how these teachers include ESE in their daily teaching practices, moulding existing policy structures to pave the way for showcasing their environmental sensibilities (see Figure 3.2). A harmonious balance with this

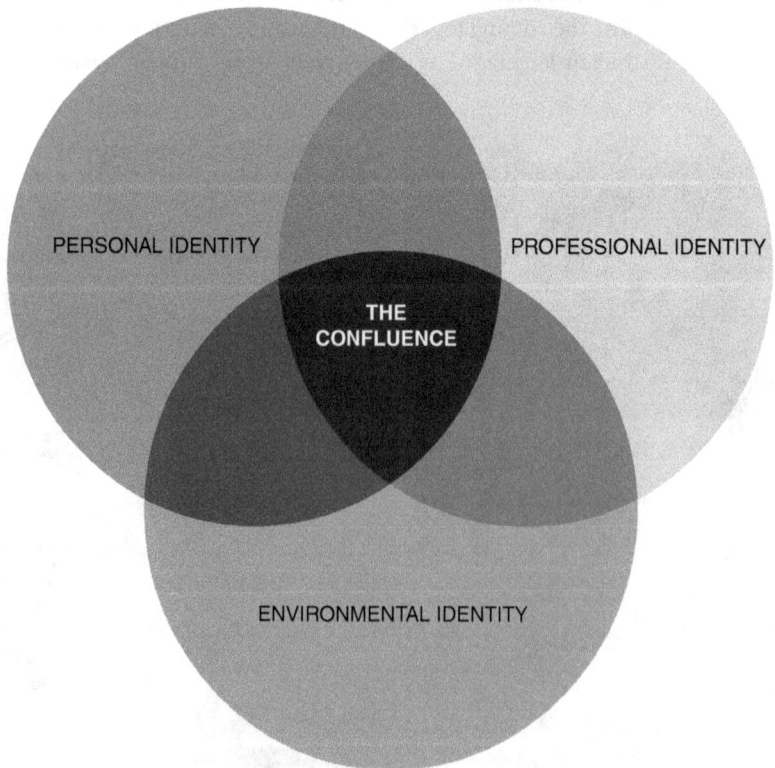

FIGURE 3.2 The confluence of identities. (Visual representations by Barnes, Moore & Almeida)

confluence can provide an individual sense of satisfaction and a feeling of making a difference in their lives as individuals, teachers and environmentalists. Identities that merge and flow synchronously allows individuals to bring their 'self' and who they are more broadly to their profession. Identities, when merged, are felt and interpreted differently, and being able to work in tune with these identities provides a strong sense of accomplishment to individuals. In contrast, over-adjusting one's beliefs, values and sense of self to meet professional, institutional or policy requirements can contribute to a dissonance and a void within individuals. Dissonance can be unsettling and create tension. However, it can also lead to further learning where an individual stretches out to find new ways towards some kind of a new balance. For example, in this study it would mean teachers find ways to negotiate around the policy constraints or institutional expectations. This means early career teachers being innovative and thinking outside the box to 'sneak in' ESE into the schools and their everyday practices. In another instance, teachers in a study conducted by Green and Somerville (2015) were innovative in finding resources by intentionally fostering community relationships. They intentionally engaged in pedagogical partnerships with wider community members to support their own and their students' understandings of sustainability by building on their environmental knowledge, skills and community heritage.

The other end of the spectrum is where these early career teachers feel stifled and harbour a deep sense of dissatisfaction, leading to attrition and causing them to rethink the workplace or, in some cases, their whole profession.

In the next chapter, we build on the theoretical construct of identity by using Bourdieu's concept of habitus to further operationalise our understanding of individuals' ways of thinking and doing as they negotiate different social contexts.

The aim of our study is to understand how early career teachers with strong environmental identities navigate the confluence and any resulting dissonance as they implement ESE in their everyday practices.

References

Akkerman, S. F., & Meijer, P. C. (2011). A dialogical approach to conceptualizing teacher identity. *Teaching and Teacher Education, 27*, 308–319.

Almeida, S. C. (2015). *Environmental education in a climate of reform*. Rotterdam, The Netherlands: Sense Publishers.

Almeida, S. C. (2020). Alternative worldviews on early childhood education for sustainability. In S. Elliott, E. Ärlemalm-Hagsér & J. Davis (Eds.), *Researching early childhood education for sustainability: Challenges, assumptions and orthodoxies* (pp. 38–50). New York, NY: Routledge.

Almeida, S., Moore, D., & Barnes, M. (2018). Teacher identities as key to environmental education for sustainability implementation: A study from Australia. *Australian Journal of Environmental Education.* doi:10.1017/aee.2018.40

Beijaard D., Meijer P., & Verloop N. (2004). Reconsidering research on teachers' professional identity. *Teaching and Teacher Education, 20*(2), 107–128.

Brenner, P. S., Serpe, R. T., & Stryke, S. (2014). The causal ordering of prominence and salience in identity theory: An empirical examination. *Social Psychology Quarterly,* 77(3), 231–252.

Burke, P. J. (2003). Relationships among multiple identities. In P. J. Burke, T. J., Owens. R. T., Serpe, & P. A. Thoits (Eds.), *Advances in identity theory and research* (pp. 195–214). New York, NY: Springer.

Clayton, S. (2012). Environmental identity: A conceptual and an operational definition. In S. Clayton (Ed.), *The Oxford handbook of environmental and conservation psychology* (pp. 45–65). Oxford: Oxford University Press.

Clayton, S., & Opotow, S. (2003). *Identity and the natural environment: The psychological significance of nature.* Cambridge, MA: MIT Press.

Clayton, S. D., & Myers, G. (2015). *Conservation psychology: Understanding and promoting human care for nature* (2nd ed.). Hoboken, NJ: Wiley Blackwell Publishers.

Connelly, F. M., & Clandinin, D. J. (1999) *Shaping a professional identity: Stories of educational practice.* New York, NY: Teachers College Press.

Dalone Stan (2009). The confluence of rivers. Retrieved on November 16, 2020 from "File:Sotočje Vipave in Hublja. Confluence of rivers Vipava and Hubelj. (3282732600). jpg" by Stan Dalone is licensed under CC BY-SA 2.0

Dauenhauer, B., & Pellauer, D. (2014). Paul Ricoeur. In E. N. Zalta, (Ed.), *The Stanford encyclopedia of philosophy* (Fall 2020 ed.). Retrieved from. https://plato.stanford.edu/archives/fall2020/entries/ricoeur/.

Day, C. (2018). Professional identity matters: Agency, emotions, and resilience. In P. A. Schutz, J. Hong & D. C. Francis (Eds.), *Research on teacher identity: Mapping challenges and innovations* (pp. 61–70). New York, NY: Springer International Publishing.

Eliasson, E. (2019). Conflicts and tensions in the constructions of vocational teacher identities. In L. M. Herrera, M. Teras & P. Gougoulakis (Eds.), *Facets and aspects of research on vocational education and training at Stockholm University: Emergent issues in research on vocational education & training* (pp. 240–267). Stockholm: Premiss forlag.

Erikson, E. H. (1956). The problem of ego-identity. *Journal of the American Psychoanalytic Association, 4,* 56–121.

Erikson, E. H. (1968). *Identity: youth, and crisis.* New York, NY: W.W. Norton.

Green, M., & Somerville, M. (2015). Sustainability education: Researching practice in primary schools. *Environmental Education Research, 21*(6), 832–845. doi:10.1080/1350 4622.2014.923382

Hall, S. (1990). Cultural identity and diaspora. In J. Rutherford (Ed.), *Identity: community, culture, difference* (pp. 222–237). London, UK: Lawrence & Wishart.

Hart, P. (2003). *Teachers' thinking in environmental education.* New York, NY: Peter Lang Publishing Inc.

Hokowhitu, B. (Ed.). (2010). *Indigenous identity and resistance: Researching the diversity of knowledge.* Dunedin, NZ: Otago University Press.

Jones, S. R., & McEwen, M. K. (2000). A conceptual model of multiple dimensions of identity. *Journal of College Student Development, 41*(4), 405–414.

Lortie, D. C. (1975). *School teacher: A sociological study.* Chicago, IL: Chicago University Press.

Meltzer, P. E. (2010). The thinker's thesaurus. (Expanded 2nd ed.) N.Y. London, UK: W.W. Norton & Co.

Miao, R. E., & Cagle, N. L. (2020). The role of gender, race, and ethnicity in environmental identity development in undergraduate student narratives. *Environmental Education Research, 26*(2), 171–188. doi:10.1080/13504622.2020.1717449

Milstein, T. (Ed.), Castro-Sotomayor, J. (Ed.). (2020). *Routledge handbook of ecocultural identity*. London, UK: Routledge. doi:10.4324/9781351068840

Ricoeur, P. (1988). *Time and narrative*, Vol. III, trans. K. Blarney and D. Pellauer. Chicago, IL: University of Chicago Press.

Ricoeur, P. (1992). *Oneself as another*, trans. K. Blamey. Chicago, IL and London, UK: University of Chicago Press.

Rodgers, C., & Scott, K. H. (2008). The development of the personal self and identity in learning to teach. In M. Cochran-Smith, S. Feiman-Nemser, J. McIntyre & K. Demers (Eds.), *Handbook of research on teacher education* (3rd ed., pp. 732–756). New York, NY: Routledge.

Ryan, R. M., & Deci, E. L. (2003). On assimilating identities to the self: A self-determination theory perspective on internalization and integrity within cultures. In M. R. Leary & J. P. Tangney (Eds.), *Handbook of self and identity* (pp. 253–272). New York, NY: Guilford Press.

Sachs, J. (2005). Teacher education and the development of professional identity: Learning to be a teacher. In P. Denicolo & M. Kompf (Eds.), *Connecting policy and practice: Challenges for teaching and learning in schools and universities* (pp. 5–21). Oxford, UK: Routledge.

Sageidet, B., Almeida, S. & Dunkley, R. (2018). Children's access to urban gardens in Norway, India and the United Kingdom. *International Journal of Environmental and Science Education, 13*(5), 467–480.

Schutz, P. A., Nichols, S. L., & Schwenke, S. (2018). Critical events, emotional episodes, and teacher attributions in the development of teacher identities. In P. A. Schutz, J. Hong & D. C. Francis (Eds.), *Research on teacher identity: Mapping challenges and innovations* (pp. 49–60). New York: Springer International Publishing.

Spencer, M. B., & Markstrom-Adams, C. (1990). Identity processes among racial and ethnic minority children in America. *Child Development, 61*(2), 290–310. doi:10.1111/j.1467-8624.1990.tb02780.x.

Stets, J. E., & Burke, P. J. (2014). The development of identity theory. *Advances in Group Processes, 31*, 57–97.

Stets, J. E., & Serpe, R. T. (2013). Identity theory. In J. DeLamater & A. Ward (Eds.), *Handbook of social psychology* (pp. 31–60). New York, NY: Springer.

Stryker, S. (1980). *Symbolic interactionism: A social structural version*. Menlo Park, CA: Benjamin-Cummings Publishing Company.

Thomashow, M. (1996). *Ecological identity: Becoming a reflective environmentalist*. Cambridge, MA: The MIT Press.

Waterman, A. S. (1985). *Identity in adolescence: Processes and contents*. San Francisco, CA: Jossey-Bass.

Werbner, P. (2017). Barefoot in Britain–yet again: On multiple identities, intersection (ality) and marginality. *The Sociological Review, 65*(1_suppl), 4–12.

Zaradez, N., Sela-Sheffy, R., & Tal, T. (2020). The identity work of environmental education teachers in Israel. *Environmental Education Research, 26*(6), 812–829. doi: 10.1080/13504622.2020.1751084

4

ENACTING AGENCY AND NEGOTIATING POWER

A theoretical framework

It would be remiss to begin this chapter without recognising the theoretical frameworks that continue to evoke and provoke meaning in the field of Environmental and Sustainability Education (ESE), as they also provide an important foundation for the theories that underpin the research presented in this book.

The positioning of humans has defined how ESE is understood and explored. Some researchers have employed systems thinking as a way to explore how humans impact and are impacted by social and biological systems (Barnes, Moore & Almeida, 2018; Elliott & Davis, 2018; Lewis, Mansfield & Baudains, 2014; Martin, Brannigan & Hall, 2007). In contrast, posthumanism approaches challenge the notion that our understanding and conceptualisations of ESE must position humans as being central to the experience and/or system. Somerville (2017) argues that posthumanism approaches in ESE 'share a focus on re-thinking the human subject as co-constituted within the more-than-human world' (p. 19). By challenging this notion that humans are central to qualitative inquiry within the field of sustainability, posthumanism confronts anthropocentric beliefs, arguing that sustainability of the environment needs to be conceptualised beyond the centralisation of human existence (Somerville, 2017; Viegas et al., 2016). Duhn (2017) highlights the tensions attached to this approach: 'The difficult task is to both decentre "the human" to generate new spaces for multispecies engagements and to take responsibility for humanity's historical attachment to human exceptionalism' (p. 47).

In addition to the use of posthumanism as a theoretical and ontological frame, the use of critical theory continues to hold a prominent place in ESE research (e.g. Brantmeier, 2013; Elliott & Davis, 2018; Kearins & Springett, 2003). Critical theorists interrogate educational knowledge and practices, exploring how they are reproduced and reconstructed in educational settings in light of issues surrounding power, inequality and marginalisation (Elliott & Davis, 2018). Critical theory provides opportunities to promote active and emancipatory learning, positioning humans as central to transformative change. For example, a number of researchers have examined how young children have become active agents in making decisions and acting on solutions to local environmental problems (Elliott & Davis, 2018; Moore, Morrissey & Robertson, 2019). Critical theory, therefore, creates an important foundation for this study's theoretical frame as we aim to explore, through Foucauldian and Bourdieuian theoretical concepts, how teachers can be active agents in promoting their ESE identities and practices within teaching contexts that often reproduce knowledge and practices that tend to downplay and marginalise ESE.

Teachers and their positioning within institutional practices

Teachers, and particularly early career teachers, routinely position and/or locate themselves within institutional discursive—normative practices (Davies & Harré, 1990; Matsunaga et al., 2020). As they negotiate institutional practices, they attempt to enact their agency as they interpret these practices, consider and imagine future actions and activate change while they 'advocate for their professional and moral beliefs and responsibilities' (Colegrove & Zuniga, 2018, p. 190). This empowerment or agency is evident in the ways in which teachers draw upon their beliefs and convictions to mitigate institutional norms and structures, which may be both explicit (e.g. policies, curriculum) and implicit (as determined by intuitional culture and ethos) (e.g. Matsunaga et al., 2020). Colegrove and Zuniga (2018) argue that:

> Yet, even in overly constrained environments, teachers enact agency. Agentic practices sometimes manifest through individual, subtle acts of resistance against imposed top-down policies perceived to be counterproductive to learner success....
>
> *(p. 190)*

Teachers have the capacity to exert power over power through alignment, resistance and/or utilisation of resources that empower them. However, Foucault extends this positioning of teachers by arguing that while power may be exercised by all, it is often extended by those who are positioned for power:

> Power is exercised rather than possessed; it is not the 'privilege', acquitted or preserved, of the dominant class, but the overall effect of its strategic positions—an effect that is manifested and sometimes extended by the position of those who are dominated. Furthermore, this power is not exercised simply as an obligation or a prohibition on those who 'do not have it'; it invests them, is transmitted by them and through them; it exerts pressure upon them, just as they themselves, in their struggle against it, resist the grip it has on them.
>
> *(Foucault, 1977, pp. 26–27)*

Schools, with their self-formed institutional identities (Bartlett, McDonald & Pini, 2015; Brickson, 2007) and their leadership structures, are positioned to exercise power by investing, exerting and transmitting power through their normative and discursive practices. Of particular relevance for early career teachers, it is important to explore how teachers perceive their capacity and/or agency as they mediate their hybrid identities (e.g. personal, professional and *environmental* identities) within contexts with well-established institutional identities. There has been a great deal written on early career teachers' feelings of inadequacy and incompetence when faced with challenging situations with learners (e.g. Berliner, 1986; Heikonen, Pietarinen, Pyhältö, Toom & Soini, 2017; Soini, Pyhältö & Pietarinen, 2010) and their feelings of isolation (Buchanan, 2012). Additionally, however, there has been a growing interest within the literature regarding the role of identity construction, particularly the mediation of personal and professional identities, which play important roles in teacher attrition (Beauchamp & Thomas, 2009; Trent, 2017). The study presented in this book attempts to capture the attitudes, beliefs and experiences of early career teachers as they negotiate institutional norms with their professional, personal and ecological identities.

Bourdieu's thinking tools: field, habitus and capital

Bourdieu's 'thinking tools' are key components of the theoretical framework underpinning this study, because these tools allow for a thorough examination of the often hidden, invisible, social relations that exist within educational settings. Bourdieu's thinking tools of field, habitus and capital will be used in this chapter to investigate not only how these concepts or tools relate to one another but how they reveal the power relations at play within particular educational settings (or fields). To begin, we will explore the concept of habitus, particularly in relation to how one's individual habitus interacts, negotiates and resists collective or institutional habitus. Both individual habitus (e.g. individual teachers) and institutional habitus (e.g. schools) are realised, shaped and/or become active in a particular field (Bourdieu, 1990a), and in the case of this study, the field of education. Habitus is characterised as 'acquired, socially constituted dispositions' (Bourdieu, 1990a, p. 13) that are unconsciously developed (Bok, 2010). Bourdieu argues that habitus shapes and is shaped by a field and is shaped and informed by both the past and present. However, 'while the habitus allows for individual agency it also predisposes individuals towards certain ways of behaving' (Reay, 2004, p. 433). Therefore, the practices and stances that are (re)produced by individual and/or institutional habitus may differ within the same field (Bourdieu, 1990a) but inevitably reflect the social context in which they are acquired (Reay, David & Ball, 2005, p. 36).

In exploring how habitus, both individual and institutional, impact early career teachers' beliefs and actions, which either lead to feelings of empowerment or that of incompetence and inadequacy, it is important to understand how habitus, capital and field relate to teacher practices and more specifically teacher agency. Reay (2004), drawing from Bourdieu's ideas, argues that 'it is through the workings of habitus that practice (agency) is linked with capital and field (structure)' (p. 432). Therefore, to understand teacher agency in enacting and implementing ESE practices in their teaching, it is important to first discuss the role of habitus and its relationship to teachers' and institutions' collective beliefs and actions. While Bourdieu argues that habitus can generate a range of possible actions and practices, often times this repertoire is confined to the social field in which habitus exists. Bourdieu (1990b) argues that the habitus reproduces 'the social conditions of our own production' (p. 87), and while habitus can result in individual actions, they are often confined to those that are socially acceptable within a particular field. In many ways, practices borne out of a particular field may be predictable:

> The habitus, as a system of dispositions to a certain practice, is an objective basis for regular modes of behaviour, and thus for the regularity of modes of practice, and if practices can be predicted... this is because the effect of the habitus is that agents who are equipped with it will behave in a certain way in certain circumstances.
>
> *(Bourdieu, 1990a, p. 77)*

While habitus may reflect the social field, it is argued that choice is still at the heart of habitus, but these choices are limited:

> I envisage habitus as a deep, interior, epicentre containing many matrices. These matrices demarcate the extent of choices available to any one individual. Choices are bounded by the framework of opportunities and constraints the person finds himself/herself in, her external circumstances. However, within Bourdieu's theoretical framework he/she is also circumscribed by an internalized framework that makes some possibilities inconceivable, others improbable and a limited range acceptable.
>
> *(Reay, 2004, p. 435)*

In other words, while habitus may be shaped by the reproduction of the social norms of the field or social context, each habitus is also shaped by its individual histories or past and therefore no two individual habitus are identical (Bourdieu, 1990b, p. 87)

Institutional habitus

The identity of an institution is often fashioned through the symbols, processes and behaviours observed by others, making them distinct from other organisations (Bartlett et al., 2015). Identity also construes how these organisations relate and position themselves in relation to their stakeholders (Bartlett et al., 2015;

Brickson, 2007). However, institutional identity suggests that their identity is solely defined by the perspectives of others.

Underpinning this study is the idea that institutional habitus plays a role in shaping, directing and/or conflicting with individual teacher habitus, and it is not something that is just observed by others but that both shapes and is shaped by the context. Drawing on the work of Bourdieu and Reay, institutional habitus is used in this book to describe the impact of a school's collective habitus, which is often shaped by the leadership of the school, on individual teacher habitus. However, just like individual habitus, institutional habitus is shaped by its own history and has been established over time (Reay et al., 2005).

Reay, Ball, David and Davies (2001) define institutional habitus as 'a complex amalgam of agency and structure and could be understood as the impact of a cultural group or social class on an individual's behaviour as it is mediated through an organization' (p. 856). Therefore, not only does the field (e.g. education) shape (while also being shaped by) individual habitus but the institutional habitus of the schools that they work within. Additionally, it is the norms, discourse practices, power relationships and rules that are accepted by a particular context that shape an institution's identity. Bourdieu refers to these contexts, shaped by norms, power relations and rules as social field, which then give 'cues and impart logic to an individual' (Dalal, 2016, p. 235). Therefore, institutional identity is shaped by the social field, which is more broadly the field of education, but is defined more specifically for each context through the establishment of explicit and implicit rules, norms and practices that have been accepted and/or legitimised by the members of the field. Early career teachers, new to the field of education, must dislocate their own identities or habitus to be able to negotiate their own beliefs and convictions with those held by their specific school in order to *play the game* within this overarching educational field. This draws upon Bourdieu's use of 'game theory' in understanding how social actors negotiate the rules of the game (Bourdieu, 1977). Through the process of negotiation, power can be exercised rather than lost, as teachers enact agency to either align their practices to match those accepted within this social field, resist or find a way to balance their alignment and resistance.

Building cultural and social capital

As early career teachers attempt to become members of a school community, they position themselves in order to build capital in the form of both cultural and social capital. While Bourdieu's concept of cultural capital refers to attaining cultural goods such as values, skills and knowledge, social capital refers to the network-based resources that are made available through relationships. Early career teachers attempt to not only move from graduate to proficient teachers, as depicted in the Australian Professional Standards for Teachers (AITSL, 2019), to build their cultural capital but they also attempt to negotiate the values and skills and knowledge that are valued by their specific school. In order for teachers to develop cultural capital, they must navigate and negotiate their habitus along-side the institutional habitus; and when these two habitus are in conflict, the individual habitus is dislocated and attempts to better align with the field in which habitus, both individual and institutional, are realised. By aligning with values, skills and knowledge base held by a particular social context, early career teachers become legitimised and more audible within this context (Miller, 1999). Additionally, while early career teachers build social capital among their new colleagues, they must balance their moral convictions, beliefs and practices with those that are socially acceptable within this new community.

In her study exploring pedagogical cultural identity among early career Aboriginal teachers, Burgess (2016) argues that education systems in the West reproduce school cultures that primarily reflect Eurocentric values, beliefs and practices and position different knowledges, cultures and peoples as the 'Other' (p. 109). She argues that pedagogical cultural identity:

> encapsulates the way in which these teachers embed their tacit cultural knowledge, passion, skills and lived experience into their daily teaching practice. In this context, conflating pedagogy, cultural knowledge, lived experience and identity became critical to an understanding of self as cultural being, teacher and learner'.
>
> *(Burgess, 2016, p. 109)*

While these teachers attempt to embed their cultural knowledge and lived experiences into their pedagogy, there are accepted norms and 'knowledges' that exert power over their teaching decisions and actions. Burgess (2016) explains how many of the early career teachers in her study drew upon their lived experiences to build capital with their learners; however, her study did not draw on how they build social capital among their colleagues, particularly in social contexts that may not reflect their lived experiences, knowledge and passions.

In another study, Kirkby, Moss and Godinho (2017) explore early career teachers' mentoring experiences in Victoria, Australia, as part of their induction programme. Kirkby et al. (2017) argue that a 'secondary habitus in the field of teaching' is characterised by: (1) the private nature of learning to teach, (2) the expectation that early career teachers can manage a workload similar to their more experienced colleagues and (3) the building of professional autonomy, which has a hidden message of competence with particular views regarding what constitutes professional knowledge (p. 24). They found that, as depicted by an early career teacher narrative, the mentor relationship was limited in that the mentor teacher lacked understanding and commitment to the role, resulting in the participant having to navigate the transition to professional autonomy on her own. She lacked the opportunity to build social capital among her colleagues but was expected to gain cultural capital in isolation.

Burgess (2016) and Kirkby et al. (2017) suggest that the habitus' of early career teachers are shaped by the field (e.g. education) and defined by their opportunities to build capital, which informs their practices. However, these two studies also suggest that while the participants are victims of the power relations that exist within their social contexts, they are active agents in tapping into their lived experiences. The participants' agency can be seen to sit alongside their present determination to ensure that they become knowledgeable and legitimate members within the field of education.

Powerful knowledge

In this book, we explore how the current field of Australian education is shaping and is shaped by 'knowledge,' by exploring what knowledge and more importantly powerful knowledge entails and how pre-service teachers must learn to interpret and negotiate the social reproduction of powerful knowledge in Australia.

Young (2009) argues that knowledge has 'become the major organising category in the educational polices of international organisation and many national governments' (Young, 2009, p. 193). With education being a key factor in the production of economic growth (Cochran-Smith et al., 2017), the Australian government mentions the integral nature of 'human capital in its long-term social and economic prosperity' (Australian Government Productivity Commission [AGPC], 2012, p. iii). More importantly, however, is determining what knowledge(s) is required to allow nations 'to keep pace in the global education race' (Barnes & Cross, 2018, p. 1). According to *Quality Schools, Quality Outcomes*, the Australian Government argues:

> education is the foundation of a skilled workforce … The better literacy and numeracy skills a young person has, the more likely they are to continue at school, undertake tertiary study, and go on to highly skilled and paid work.
>
> *(AGDET, 2016, p. 1)*

Young (2009) argues that in many policy documents, there is a global trend of positioning 'knowledge' as abstract, vague and devoid of content. However, with a purposeful focus on literacy and numeracy skills in recent education policy documents (Australian Government Department of Education & Training [AGDET], 2016, 2018), *knowledge* in Australia is shaped by a disciplinary focus on literacy and numeracy skills. So much so that since 2016, pre-service teachers in Australia are required to pass a literacy and numeracy test after acceptance into their education programme to further demonstrate that they have the required literacy and numeracy skills to be a classroom teacher (Barnes & Cross, 2018). However, even within this concept of literacy, there are aspects that are not considered as socially and educationally valuable as others. In the context of English language development among migrants and refugees, Miller and Windle (2010) argue that 'language work in the content classroom is given little status when set alongside other knowledge hierarchies supported by societal and educational agendas' (p. 34). The existence of knowledge hierarchies causes pause when considering how ESE is positioned, even in the midst of aspirational, yet not operational, educational policy (e.g. AGDET, 2018; Ministerial Council on Education, Employment, Training and Youth Affairs, 2008).

Given that the word *knowledge* has both a public and historical association with the word *truth* (Foucault, 1977; Young, 2009), the knowledge(s) that is needed to be an educated and highly skilled contributor to a nation's workforce has a problematic place within power relations. Who makes the decisions when considering the hierarchy of knowledge(s)? How is this shaped by different time periods and different societies? How is this hierarchy of knowledge operationalised in a meaningful curriculum that can be used in a range of different contexts? Foucault (1977) argues that:

> power produces knowledge (and not simply by encouraging it because it is useful); that power and knowledge directly imply one another; that there is no power relations without the correlative constitution of a field of knowledge, nor any knowledge that does not presuppose and constitute at the same time power relations.
>
> *(p. 237)*

In other words, you cannot separate power and knowledge because they are inextricably linked. Foucault (1977) further argues that knowledge is not defined by a dichotomy of either being resistant or useful to power but is inseparable, and it is 'the processes and struggles that traverse it [power-knowledge] and of which it is made up, that determines the forms and possible domains of knowledge.' Just as Bourdieu (1990a) demonstrates that power relations are at play in a particular field, and individual and institutional habitus are shaped and negotiated within this field, knowledge is also shaped and negotiated and determined by the powerful players and social structures that organise any given field.

This then gives light to the role of curriculum design in defining what knowledge is deemed important in a particular field. Young and Muller (2013) argue that formal education requires a knowledge-based curriculum and that *specialist* knowledge is powerful knowledge. They argue further that: 'Powerful knowledge is powerful because it provides the best understanding of the natural and social worlds that we have and helps us go beyond our individual experiences' (p. 196). However, a focus on specialist knowledge has been at odds with recent international curricular trends, particularly with the creation of national curriculum that focuses on outcomes, 'generic' skills, capacities or competencies and cross-cutting or transversal themes (Priestly, 2011; Tedesco, Opertti & Amadio, 2014). Priestly (2011) argues that there is a removal of specific and detailed knowledge and content in these new models. Therefore, on one hand, it becomes difficult to distinguish what knowledge is important for human capital building if it is not detailed and specific, but on the other hand, if decisions are made as to what knowledge is more important, whose voices are heard?

Van Manen (1990) suggests that knowledge forms are not always best packaged as an external gift from someone/thing more powerful (e.g. a national curriculum board) but may be part of an internal process that teachers go through as they take agentive steps to embody knowledge through their own thoughtfulness and practice:

> The ultimate success of teaching actually may rely importantly on the "knowledge" forms that inhere in practical actions, in an embodied thoughtfulness, and in the personal space, mood and relational atmosphere in which teachers find themselves with their students. The curricular thoughtfulness that good teachers learn to display towards children may depend precisely upon the internalized values, embodied qualities, thoughtful habits that constitute virtues of teaching.
>
> *(p. 13)*

However, as the research in this book will explore, the development of a beginning teachers' individual identity (Burgess, 2016) and their embrace of their own values and embodied knowledge often comes second to more powerful knowledge that is determined by social and educational agendas. This book seeks to understand how ESE is positioned within this knowledge hierarchy and provide ways forward for teachers to take agentive steps to negotiate their individual habitus within institutional norms.

References

Australian Government Department of Education and Training. (2016). *Quality schools, quality outcomes*. Canberra: AGDET. Retrieved from: https://docs.education.gov.au/system/files/doc/other/quality_schools_acc.pdf

Australian Government Department of Education and Training. (2017). *Literacy and numeracy test for initial teacher education students: Assessment framework*. Retrieved from: https://teacheredtest.acer.edu.au/files/Literacy-and-Numeracy-Test-for-Initial-Teacher-Education-Students-Assessment-Framework.pdf

Australian Government Department of Education and Training. (2018). *Through growth to achievement: Report of the review to achieve educational excellence in Australian schools*. Retrieved from: https://docs.education.gov.au/node/50516

Australian Government Productivity Commission. (2012). *Schools workforce: Productivity Commission research report*. Retrieved from: http://www.pc.gov.au/inquiries/completed/education-workforce-schools/report/schools-workforce.pdf

Australian Institute for Teaching and School Leadership (AITSL). (2019). Australian professional standards for teachers. Retrieved from: https://www.aitsl.edu.au/teach/standards

Barnes, M., & Cross, R. (2018). 'Quality' at a cost: The politics of teacher education in Australia. *Critical Studies in Education*. doi:10.1080/17508487.2018.1558410

Barnes, M., Moore, D., & Almeida, S. (2018). Sustainability in Australian schools: A cross-curriculum priority? *Prospects, 48*(1), 1–16. doi:10.1007/s11125-018-9437-x

Bartlett, J., McDonald, P., & Pini, B. (2015). Identity orientation and stakeholder engagement: The corporatisation of elite schools. *Journal of Public Affairs, 15*(2), 201–209.

Berliner, D. (1986). In pursuit of the expert pedagogue. *Educational Researcher, 15*(7), 5–13.

Beauchamp, C., & Thomas, L. (2009). Understanding teacher identity: An overview of issues in the literature and implications for teacher education. *Cambridge Journal of Education, 39*(2), 175–189.

Bok, J. (2010). The capacity to aspire to higher education: 'It's like making them do a play without a script.' *Critical Studies in Education, 51*(2), 163–178.

Bourdieu, P. (1977). *Outline of a theory of practice.* Cambridge: Cambridge University Press.

Bourdieu, P. (1990a). *In other words: Essays towards a reflexive sociology.* Cambridge, UK: Polity Press.

Bourdieu, P. (1990b). *The logic of practice.* Cambridge, UK: Polity Press.

Brantmeier, E. (2013). Toward a critical peace education for sustainability. *Journal of Peace Education, 10*(3), 242–258.

Brickson, S. (2007). Organizational identity orientation: The genesis of the role of the firm and distinct forms of social value. *Academy of Management Review, 32*(3), 864–888.

Buchanan, J. (2012). Sustainability education and teacher education: Finding a natural habitat? *Australian Journal of Environmental Education, 28*(2), 108–124.

Burgess, C. (2016). Conceptualising a pedagogical cultural identity through the narrative construction of early career Aboriginal teachers' professional identities. *Teaching and Teacher Education, 58*, 109–118.

Cochran-Smith, M., Baker, M., Burton, S. et al. (2017). The accountability era in US teacher education: Looking back, looking forward. *European Journal of Teacher Education, 40*(5), 572–588. doi:10.1080/02619768.2017.1385061

Colegrove, K., & Zuniga, C. (2018). Finding and enacting agency: An elementary ESL teacher's perception of teaching and learning in the era of standardised testing. *International Multilingual Research Journal, 12*(3), 188–202.

Dalal, J. (2016). Pierre Bourdieu: The sociologist of education. *Contemporary Education Dialogue, 13*(2), 231–250.

Davies, B., & Harré, R. (1990). Positioning: The discursive production of selves. *Journal for the Theory of Social Behaviour, 20*(1), 43–63.

Duhn, I. (2017). Cosmopolitics of place: Towards urban multispecies living in precarious times. In K. Malone, S. Truong, & T. Gray (Eds.), *Reimaging sustainability in precarious times* (pp. 45–57). Singapore: Springer.

Elliott, S., & Davis, J. (2018). Moving forward from the margins: Education for sustainability in Australian early childhood contexts. In G. Reis & J. Scott (Eds.), *International perspectives on theory and practice of environmental education: A Reader* (pp. 164–178). Cham, Switzerland: Springer.

Foucault, M. (1977). *Discipline and punish: The birth of the prison.* New York, NY: Pantheon Books.

Heikonen, L., Pietarinen, J., Pyhältö, K., Toom, A., & Soini, T. (2017). Early career teachers' sense of professional agency in the classroom: Associations with turnover intentions and perceived inadequacy in teacher-student interactions. *Asia-Pacific Journal of Teacher Education, 45*(3), 250–266.

Kearins, K., & Springett, D. (2003). Educating for sustainability: Developing critical skills. *Journal of Management Education, 27*(2), 188–204.

Kirkby, J., Moss, J., & Godinho, S. (2017). The devil is in the detail: Bourdieu and teachers' early career learning. *International Journal of Mentoring and Coaching in Education, 6*(1), 19–33.

Lewis, E., Mansfield, C., & Baudains, C. (2014). Ten tonne plan: Education for sustainability from a whole systems thinking perspective. *Applied Environmental Education & Communication, 13*(2), 128–141.

Martin, S., Brannigan, J. & Hall, A. (2005) Sustainability, systems thinking and professional practice. *Journal of Geography in Higher Education, 29*(1), 79–89. doi:10.1080/03098260500030389

Matsunaga, K., Barnes, M., & Saito, E. (2020). Exploring, negotiating and responding: International students' experiences of group work at Australian universities. *Higher Education*. doi:10.1007/s10734-020-00592-5

MCEETYA [Ministerial Council on Education, Employment, Training and Youth Affairs]. (2008). *Melbourne declaration on educational goals for young Australians*. Retrieved from http://www.curriculum.edu.au/verve/_resources/National_Declaration_on_ the_Educational_Goals_for_Young_Australians.pdf

Miller, J. (1999). Becoming audible: Social identity and second language use. *Journal of Intercultural Studies, 20*(2), 149–165. doi:10.1080/07256868.1999.9963477

Miller, J., & Windle, J. (2010). Second language literacy: Putting high needs ESL learners in the frame. *English in Australia, 45*(3), 31–40.

Moore, D., Morrissey, A-M., & Robertson, N. (2019). 'I feel like I am getting sad there': Early childhood outdoor playspaces as places for children's wellbeing. *Early Child Development and Care*. doi:10.1080/03004430.2019.1651306

Priestly, M. (2011). Whatever happened to curriculum theory? Critical realism and curriculum change. *Pedagogy, Culture & Society, 19*(2), 221–237.

Reay, D. (2004). 'It's all becoming a habitus': Beyond the habitual use of habitus in educational research. *British Journal of Sociology of Education, 24*(4), 431–444.

Reay, D., Ball, S., David, M. & Davies, J. (2001). Choices of degree or degrees of choice? Social class, race and the higher education choice process. *Sociology, 35*(4), 855–874.

Reay, D., David, M., & Ball, S. (2005). *Degrees of choice: Social class, race and gender in higher education*. Stoke-on-Trent: Trentham Books.

Soini, T., Pyhältö, K., & Pietarinen, J. (2010). Pedagogical well-being: Reflecting learning and well-being in teachers' work. *Teachers and Teaching: Theory and Practice, 16*(6), 735–751.

Somerville, M. (2017). The anthropocene's call to educational research. In K. Malone, S. Truong, & T. Gray (Eds.), *Reimaging sustainability in precarious times* (pp. 45–57). Singapore: Springer.

Tedesco, J. C., Opertti, R., & Amadio, M. (2014). The curriculum debate: Why it is important today. *Prospects, 44*, 527–546.

Trent, J. (2017). Discourse, agency and teacher attrition: Exploring stories to live by amongst former early career English language teachers in Hong Kong. *Research Papers in Education, 32*(1), 84–105.

van Manen, M. (1990). *Researching lived experience. Human science for an action sensitive pedagogy*. Ontario, Canada: SUNY Press.

Viegas, C., Bond, A., Vaz, C., Borchardt, M., Pereira, G., Selig, P., & Varvakis, G. (2016). Critical attributes of sustainability in higher education: A categorisation from literature review. *Journal of Cleaner Production, 126*, 260–276. doi:10.1016/j.jclepro.2016.02.106

Young, M. (2009). Education, globalisation and the 'voice of knowledge.' *Journal of Education and Work, 22*(3), 193–204.

Young, M., & Muller, J. (2013). On the powers of powerful knowledge. *Review of Education, 1*(3), 229–250.

5

EMPOWERMENT THROUGH STORYTELLING

A combinational methodology

Introduction: ontological and epistemological understandings

In this chapter we will discuss the methodological design for the study. Primarily, this includes the choice of a combinational methodology blending 'research by design' with an adapted 'narrative inquiry,' the recruitment of participants and the inclusion of community partners and the methods used to capture oral and written narratives throughout the research.

However, before moving onto the methodological decisions in this chapter, it is important to examine the ontological and epistemological framework that underpins the study to illustrate how we, as researchers and authors, understand our collective world view and the nature of knowledge. As interpretative rather than objectivist researchers (Denzin & Lincoln, 2008), we believe that knowledge is a social construction of shared meanings. Therefore, the multiple meanings of the participants' and researchers' stories of their lived experiences were co-constructed, interpreted and shared together rather than simply 'out there' waiting to be discovered (Hollingsworth & Dybdahl, 2007). Further to this, Nolan, Macfarlane and Cartmel (2013) suggest this epistemological 'process allows for different constructions of meaning from different people, not one true way...' (p. 93), as the possibility of only one 'true' knowledge or truth is no longer considered achievable, nor desirable (Denzin & Lincoln, 2008). Given these understandings, the decision to use a combinational methodological approach was twofold as the methodology and associated methods needed to:

1. Provide a time and space to hear the often marginalised personal stories of early career teachers (that is, through the use of narrative inquiry);
2. Explore the participants' experiences and attempts to implement Environmental and Sustainability Education (ESE) into their everyday teaching

practices, and co-construct possible solutions to the educational 'problem' with the researchers and community partners (that is, through the principles of research by design).

The aims of the study were to enable the participants and the researchers to listen, discuss, interpret, reflect and work together while collectively attempting to make sense of the early career teachers' ESE experiences in their respective early childhood or primary classrooms. As a consequence, the research questions were:

> What role do the personal, professional and environmental identities of early career teachers play in the implementation of ESE in their everyday teaching practices?

> How do early career teachers negotiate their personal, professional and environmental identities in the midst of institutional habitus?

To fulfil the aims of the study while working towards answering these research questions, an organic, evolving methodology combining narrative inquiry within a research by design approach was adopted as the overarching methodology.

Combinational methodology: a narrative inquiry embedded within research by design

As we have explained in Chapter 2, the educational 'problem' the study was focused on was that ESE was considered to be unnecessary, too difficult and/ or too alarming to enact in early childhood and primary classrooms. Given the apparent resistance to change and inability to shift teachers' attitudes around ESE (Davis & Elliott, 2009), a new way of approaching, examining and analysing the 'problem' was sought through the use of a more innovative, combinational blend of methodologies for this study.

A methodological framework combining both an adapted narrative inquiry with research by design was deemed to have a 'clear fitness for purpose' (Cohen, Manion & Morrision, 2011, p. 115) for this study, which sought to examine the in-depth meanings of the stories told by early career teachers about their lived experiences of teaching ESE. Both narrative inquiry (Clandinin, 2013; Clandinin & Connelly, 2000) and research by design (Plomp, 2013) methodologies have become increasingly valued for their in-depth examination of complex educational issues involving teachers, teaching and learning. Many of the tenets embedded within a narrative inquiry align closely with research by design principles, and vice versa. For example, a research by design study allows for the 'research problem' to be explored, refined and adapted in a real world setting by the teachers and the researchers working closely together (Cotton, Lockyer & Brickell, 2009). Similarly, a narrative inquiry encourages a close research relationship between the researcher and the participants, where all voices are considered an integral element of the co-constructed research narratives. In both methodologies, contextual considerations are of vital significance to the participants, the analysis and the findings (Clandinin, 2013; Plomp, 2013). The following two sections of this chapter provide a brief overview of both methodological approaches, first research by design and then followed by narrative inquiry, to explain their key principles and to highlight their relevance to the study.

Research by design methodology

As an emergent, contemporary research paradigm from the 1990s (Cotton et al., 2009), research by design has become known for its capacity to 'address educational practice…in naturalistic contexts' rather than simply collecting de-contextualised data (Plomp, 2013, pp. 11–12). Plomp's (2013) definition of the methodological approach is valuable here in its comprehensive explanation of the processes involved in his research by design model, stating:

> To design and develop an intervention (such as programs, teaching-learning strategies and materials, products and systems) as a solution to a complex educational problem as well as to advance our knowledge about the characteristics of these interventions and the processes to design and develop them…
>
> *(p. 15)*

Further to this, Plomp (2013, p. 19) explains what is involved in these processes by identifying three systematic phases embedded within the model's processes as follows:

- **Phase One**: Preliminary phase—which involves a needs and/or context analysis of the 'educational problem,' a review of the literature surrounding the issue and the development of a theoretical framework;
- **Phase Two**: Development phase—the iterative design phase in each cycle of the research, showing how to create, improve and refine the interventions;
- **Phase Three**: Assessment phase—during which reflection and an evaluation is made to ascertain whether the intervention or solution has been reached, or if further improvement of interventions is recommended to cycle back to phase one and two.

Similarly, Reeves (2006) created a design-based research model which was based on a flow chart presentation that worked through the processes from the initial analysis of the 'problem' through to 'iterative cycles of testing and refinement of solutions in practice' and concluded with a reflection on the implementation of the solution (p. 59).

The following model in Figure 5.1 draws on Plomp's (2013, p. 17) definition and descriptive model of research by design, together with Reeves' (2006, p. 59) flow chart model, to illustrate how these phases could work iteratively towards solving an 'educational problem' using this methodological approach:

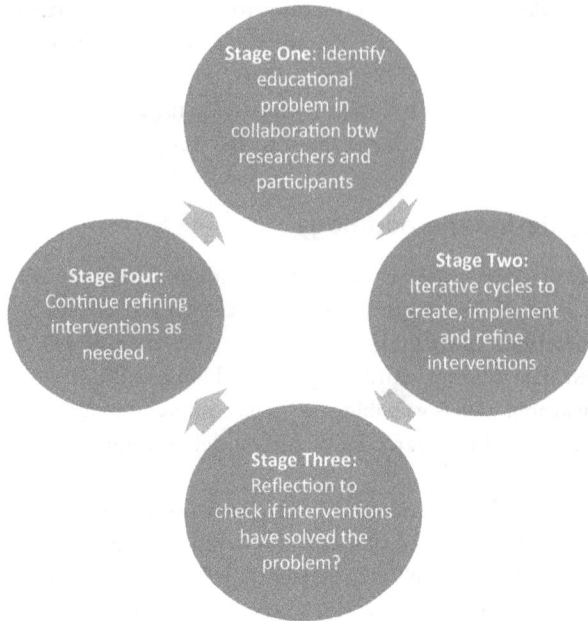

FIGURE 5.1 A visual representation of the key features of a research by design methodology. (Plomp, 2013; Reeves, 2006)

As the model in Figure 5.1 illustrates, the iterative nature of the phases of a research by design methodology are particularly noteworthy, as these phases can be repeated in a cyclical pattern building on the previous iteration until a solution is reached. With the view towards finding a solution to the education problem, the interventions are created, re-created and refined in collaboration with the participants within their context. Other researchers who have implemented a research by design methodology have found its focus on 'support for teachers' in their attempts to solve 'broad based complex, real world problems that are critical to education' as its key point of difference in contrast to more traditional research methodologies (Cotton et al., 2009, p. 1365). Further to this, these researchers argue that the continual interaction between the researchers and the participants, and the way that knowledge is constructed throughout the research process, are both points of difference within this methodology (Cotton et al., 2009).

These methodological processes, such as the interaction between the researchers and participants, have also been clearly explained in Wang and Hannafin's work to enable others (see for example Plomp's (2013) and Reeves' (2006) interpretation of the approach) to follow through similar adaptations of the methodology. Of particular relevance to our study, Wang and Hannafin (2005) highlight the key characteristics of this methodology as being 'integrative' allowing for 'mixed research methods' throughout the research process as new needs and issues emerge (p. 8); and, 'interactive, iterative and flexible' as the researchers and participants 'seek to improve an initial design plan through implementation' (pp. 9–10). While research by design provided the framework and the processes for the study, it was the principles and practices of a narrative inquiry that provided the storied content and the deeply analytical approach to the research.

Narrative inquiry methodology

Narrative research is based on the study of narratives or stories (Riessman, 2008). Kramp (2004) explains the significance of stories in our lives, and indeed to research, by stating:

> Respect for stories and appreciation for their value have grown as we have come to understand more fully how they assist humans to make life experiences meaningful. Stories preserve our memories, prompt our reflections, connect us with our past and present, and assist us to envision our future.
>
> *(p. 106)*

Expanding on Kramp's definition of narrative research, a narrative inquiry can be defined as the 'study of experience as story' (Connelly & Clandinin, 2006, p. 477). As we are all storytellers, our lived experiences are defined in the stories we choose to tell others. It is in the telling of these stories that the meanings of our experiences become clear to ourselves and to others (Clandinin & Connelly, 2000). A narrative inquiry can be seen to be different from other forms of narrative research in the analytical way researchers and participants examine 'the underlying insights and assumptions that the story illustrates' (Bell, 2002, p. 208) rather than simply report the 'stories' told through data generation. The significance of context in research is magnified through the use of narrative inquiry as a methodological approach. Clandinin and Rosiek's (2007) explanation of a narrative inquiry highlights the importance of context when they state:

> The focus of narrative inquiry is not only on individuals' experiences but also on social, cultural, and institutional narratives within which individuals' experiences are constituted, shaped, expressed and enacted.
>
> *(p. 42)*

In Clandinin and Rosiek's (2007) explanation, close links and synergies can be seen with the previous theoretical chapter on Bourdieu's theory of habitus and how the research participants' confluence of identities and teaching practices have been impacted and influenced by their contextual position. Through this combined methodological approach, context can be seen to highly influence the content of the stories the participants told. Of further relevance to the current research, are the following four key principles of a narrative inquiry:

1. The acceptance of the trustworthiness of the participants' stories at that time and place, with the understanding that all stories change over time (Riessman, 2008);

2. The 'simultaneous examination of temporal, societal and place-based dimensions' of the participants' experiences told through stories (Connelly & Clandinin, 2006, p. 479);
3. The use of an informal, conversational form of storytelling to generate research data (Clandinin & Connelly, 2000) and similar to the research by design principle;
4. The development of a researcher-participant research relationship, based on developing trust and evolving understanding (Clandinin, 2013).

A noteworthy quality of narrative inquiry is its capacity to enable the participants to 'question what they thought they knew' about a phenomenon (Golombek & Johnson, 2004, p. 306). Through this process of iterative and deepening questioning, participants were able to push back and disrupt normative ideas of early career teaching to create new insights from the close examination and analysis of the stories the participants told (Bamberg & Andrews, 2004).

The use of this combinational methodology, complexities around human behaviour and the 'small story' nuances of the participants' experiences (Clandinin & Connelly, 2000) were captured in a way that objective research methodologies would not be able to achieve. By providing space and time to hear the participants' stories over an extended period during their first year of teaching, the researchers were able to listen to and witness the shifts the participants were making in their thinking about teaching and, specifically, in teaching ESE. Initially, the participants told the 'big stories' (Bamberg & Andrews, 2004) or 'public narratives' (Somers, 1994) of teaching they had initially believed to be true for all teachers. However, over time their stories exhibited an evolving understanding that their own individual 'small stories' were of more significance to themselves and to the study. The combination of these two complementary methodologies enabled an in-depth insight into the multiple meanings of the participants' experiences whilst providing a focused, storied and iterative approach to the data gathering methods, the analytical processes and the findings subsequently identified.

Three-dimensional narrative inquiry analysis

A key element of a narrative inquiry that heavily draws on Clandinin and Connelly's (2000) pivotal work is the use of a three-dimensional analysis of the participants' experiential stories. Clandinin and Connelly (2000) explain their three-dimensional inquiry model as:

> ...our terms are personal and social (interaction); past, present, and future (continuity); combined with the notion of place (situation). This set of terms creates a metaphorical three-dimensional narrative inquiry space, with temporality along one dimension, the personal and the social along a second dimension, and place along a third. Using this 9set of terms, any particular inquiry is defined by this three-dimensional space: studies have

temporal dimensions and address temporal matters; they focus on the personal and the social in a balance appropriate to the inquiry; and they occur in specific places or sequences of places.

(p. 50)

In Clandinin and Connelly's narrative inquiry processes, the three dimensions of temporality, sociality and place are 'simultaneously examined' in each story to enable the identification of context-rich detail (Connelly & Clandinin, 2006, p. 479). Rather than choosing single words to create decontextualised categories for analysis, as is common in other forms of narrative research, the whole context of each participants' stories are examined through the three dimensions of analysis. In line with Clandinin and Connelly's (2000) three-dimensional narrative inquiry model, the analytical 'checkpoints' for this study were as follows:

1. Temporality—past, present and future experiences and aspects of the participating early career teachers and teaching of ESE in their specific early childhood and primary school contexts;
2. Sociality—descriptions of the meanings of personal and societal experiences which have informed and influenced the early career teachers and their teaching of ESE in early childhood and primary school contexts;
3. Place—particular places such as the early childhood and primary school locations (for example, rural, suburban or inner city and resources (for example, natural play-spaces, inside classrooms, local reserves, beaches, farms) informed and influenced the stories the early career teachers told about their experiences of teaching ESE.

To start the analytical process for this study, the audio-recorded transcripts from each of the workshops were closely read and reread a number of times. In an initial sweep of inductive analysis, each of the participants' stories within each of the workshops were examined closely to identify thematic themes, narrative patterns, tensions and/or gaps in the stories that were told. At first, we had thought the stories were illustrating patterns around individual teachers' characteristics. However, on further analysis of the collective storied data, we became increasingly aware that it was the context (rather than the individual) that generated the characteristics of the key themes. For example, clear patterns were evident around the existence of a 'hierarchical' school where any attempt by early career teachers to introduce ESE were 'squashed' if permission had not been given; and again, other stories illustrated the common occurrence of the 'status quo' schools where numeracy and literacy curriculum demands always trumped the early career teachers' desire to introduce ESE into their classrooms. Once a 'context' had been determined through further deductive analysis, then the relevant teachers' stories in that particular context were analysed within the three-dimensional model to ascertain the temporal, societal and place-based influences on their storied experiences in that setting. While the 'simultaneous examination' of the "head56 through storytelling

dimensions frequently overlapped in each of the stories, we found the purposeful use of this analytical framework helpful to make sense of the storied data. This final level of the three-dimensional analysis enabled a more in-depth identification of themes, patterns, tensions and/or gaps in each of the participants' stories and, as such, were the foundational framework of the narrative chapters retold in this book.

Methods: participants

The recruitment process chosen for this study was based on the concept of 'intensity sampling' (Sherwood & Reifel, 2010, p. 325). Creswell (2007) explained intensity sampling of participants as the selection of individuals who already 'had an understanding of the phenomenon' under study (p. 523). Since the aim was to examine the experiences of early career teachers who had specifically shown enthusiasm for ESE during their initial teacher education course, a random sample of participants would not have been appropriate. The researchers had taught the participants as pre-service teachers in the previous year at a university in Melbourne, Victoria, and knew of their strong personal commitment to sustainability. As the participants had now graduated and held teaching positions in either a primary or early childhood educational setting, any perceived issue around potentially unequal power relationships in a teacher/student interaction was no longer applicable.

Following ethics approval for the study, a number of newly graduated Early Years teachers (teaching birth to eight-year-old children) were contacted by email to first garner their interest and subsequently invite their participation in the research. At the beginning of the study, 11 participants enthusiastically gave their consent to participate in the study; nine primary school teachers (six females, three males) and two early childhood teachers (two females). The 11 participants were invited to attend a series of workshops spread over seven months (May to November) when convenient with their teaching commitments. Research monies were allocated towards Casual Relief Teaching (CRT) payments for each of the participants' school or early childhood service to allow the teachers to attend the workshops during work time. This was an ethical decision to raise the profile of the teachers and the research rather than expecting the teachers to participate after work. The initial research design intention was that the participants would be inspired and encouraged through a 'community of practice' environment to create and work on individual ESE projects within their educational settings.

Methods: setting up a 'community of practice' environment

As part of the preliminary phase of the research, the setup of a community of practice environment was needed where participants would be invited to tell stories of their past, present and/or future experiences of ESE in their respective educational settings. Jorgensen, Edwards and Ipsen (2018) describe a 'community of practice' as a:

> group of people who share a concern, a set of problems, or a passion about a topic, and who deepen their knowledge and expertise in this area by interacting on an ongoing basis.
>
> *(Wenger et al., 2002, cited in Jorgensen et al., 2018, p. 1030)*

This definition of a 'community of practice' aligns well with the purpose of the research design where a series of five workshops over seven months were conducted with the participants, researchers and representatives from two sustainability focused community partnership organisations, the[1] Dolphin Research Institute (DRI) and the Centre for Education and Research in Environmental Strategies (CERES). The DRI (see for more information *DRI – Dolphin Research*) is based in Hastings in Western Port Bay in Victoria, and has at its core a commitment to research and education around marine conservation. The DRI work includes citizen science projects (such as reporting sightings of whales and dolphins in the Two Bays); educational programmes (such as *i sea, i care* Ambassador programmes in primary schools); and environmental leadership programmes (such as pollution control in marine environments). Similarly, the CERES (see for more information *About CERES*), based in East Brunswick, inner Melbourne on the Merri Creek, focuses on community-based learning and action that is environmentally beneficial and socially just, to name a few of their key goals. Sustainable school workshops and support through *ResourceSmart* programmes are provided as well as incursions or excursions such as climate crisis, changing behaviours or the four pillars of sustainability (water, waste, biodiversity and energy) that are part of their suite of educational workshops. Previously involved in educational programmes with the researchers, both of the representatives from each of these community partner organisations were keen to be involved as active partners in the research.

Another key purpose of the community of practice workshops draws on Bhabha's (1994) notion of a 'third space,' to create an impartial meeting place away from the participants' educational settings for our conversational storytelling. As a consequence, this space could be seen as a safe meeting place for the participants to tell and re-tell their stories of their attempts, challenges and shifts in thinking about ESE without the fear of being interrupted or ridiculed by others. Within the safe 'third space' of the workshops, it was anticipated that with the inspirational ideas and support from the community partners and the researchers, the participants would be motivated to create their own ESE project to be introduced into their early childhood or primary school classrooms. In creating this community of practice environment, it was also anticipated that the early career teachers would also feel supported by their small group of fellow research participants as they worked through the issues and barriers surrounding early career teaching as well as ESE experiences.

Methods: conversational storytelling during the workshops

During each workshop, the participants were invited to engage in conversational storytelling (Clandinin & Connelly, 2000) about their lived experiences of attempting to implement ESE in their educational setting. The community partners took alternate turns in providing advice, skills and resources to participants as prompts and inspiration for potential ESE experiences in the classroom. In line with narrative inquiry tenets, the researchers also contributed to the conversations with their own storied experiences to continue building a trusting relationship between the participants and the researchers. Research data were generated through audio-recorded conversational storytelling between the participants, the researchers and the community partners, with permission given for the audio-recordings to be recorded and transcribed. Research artefacts as data were also generated through a variety of templates, resource materials and physical prompts provided by the researchers and community partners as part of their presentations.

The following diagram, Figure 5.2, illustrates the series of six workshops that were conducted over the seven-month period of the study encompassing the basis for the research by design framework:

A key element of the preliminary phase of the research was the identification of the educational 'problem' we aimed to work on in the study (see Figure 5.3).

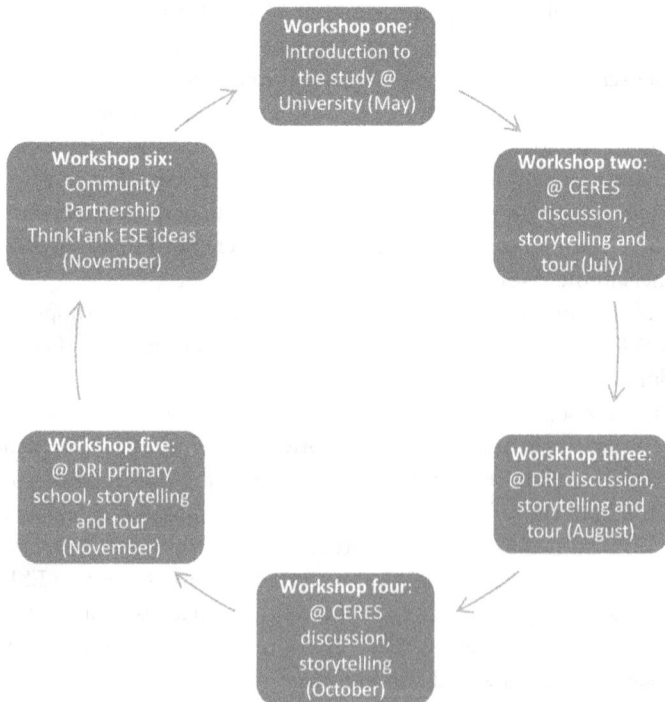

FIGURE 5.2 Visual representation of the six workshops. (Visual representations by Barnes, Moore & Almeida)

This was initially done through an analysis of literature on the global environmental crisis and urgent need for a more targeted approach to ESE in the early years of education. As a consequence, the project was developed into an iterative series of community of practice workshops to discuss and examine this 'problem' and how it could be addressed by the early career teachers in their educational settings. Each workshop within the organic research by design framework had its own proposed agenda based on content informed by the proceeding workshop. This collaboration was evident after the first workshop, whereby the subsequent workshops were co-constructed by the early career teachers, the community partners and the researchers. An intrinsic part of the collaboration was through questioning, reflective evaluations and refined interventions that were redeveloped on a regular basis as part of the research approach.

Workshop one: introduction to the study

The introductory workshop one was held in the university science room, the site where the pre-service teachers had been initially introduced to ESE and to provoke memories and stories of their passion for sustainability. During this workshop, the participants were given time to talk through their experiences as pre-service teachers, with prompting questions such as:

What were the most meaningful aspects of pre-service ESE? Why?

How have these pre-service experiences influenced what you think and do in the classroom now?

Although within a more formal questioning format than is common in a narrative inquiry (Clandinin & Connelly, 2000), the questions were used as a trigger for individual and group conversational storytelling rather than requiring 'one correct' response (Riessman, 2008). Following on from this reflective storytelling from the past, the participants were then invited to think about their current teaching context, with question prompts such as:

How would you describe your experience as a teacher trying to implement ESE?

The first workshop concluded with a 'looking forward' segment, where the participants were asked to deeply reflect:

Now that you know what you know now, what is your advice for us as teacher educators to better prepare pre-service teachers for the realities of the classroom with regard to ESE?

An action plan of potential ESE intentional teaching was requested at the conclusion of this first workshop, inviting the participants to shape an 'organic research project' with a wide variety of ideas they could enact in their educational settings. Although enthusiastic, not one participant offered any suggestions nor filled out an action plan at this stage of the project. An online blog was introduced at this early stage for participants to contribute ideas, questions and thoughts about their ESE experiences as an adjunct to the community of practice workshops (see Figure 5.3).

Workshop two: CERES conversations and tour

At the start of each proceeding workshop, time and space were given to the participants to talk through ideas, stories and questions as well as any follow-up questions from the previous workshop. For example, in workshop two, the prompt for further storytelling was:

> How do you (or plan to) balance your interest/passion for ESE with your obligations and responsibilities as a teacher?

> How "doable" do you feel creating and implementing an action plan is?

> How are you feeling about this project? Excited? Frustrated? Obligated?

Indeed, this storytelling time became such a crucial focus of the study, with the participants so eager to talk through their storied experiences with each other, the researchers and the community partners that an increasing amount of dedicated time at the beginning of each workshop was needed for this experience. In line with the second phase on the development of a research by design project, any 'interventions' that are introduced may need to be improved and refined over time. Consequently, the time allocation for each section of the workshop changed considerably to accommodate much more storytelling. This was an early refinement of the intervention to the workshop format that had not been anticipated (see Figure 5.3).

Each workshop had a different focus to follow on from the proceeding conversational storytelling sessions. For example, the second workshop was held at CERES, with the CERES community partner presenting behaviour change advice on how to 'start a buzz' about a sustainability issue in an educational setting amongst the children, families, other teachers and educational leaders. Suggestions starting from 'creating a conversation' and 'building ownership' to finally making an official start to an ESE project were offered and discussed in detail. Following on from the CERES community partner's discussion, a walking tour of the CERES environmental park provided additional time and space for the participants to further reflect and question their assumptions about teaching ESE as early career teachers and the challenges they faced. As other researchers have found (cf: Kuntz & Presnall, 2012), talking whilst walking side by side stimulated in-depth storytelling amongst the participants, researchers and community partners and became another significant element of this study that had not been foreseen.

The agenda for workshop two included a written request to 'Complete the action plan and send to the researchers as soon as possible'; however, again no action plan was forthcoming. It should also be noted that none of the initial prompt questions around action plans were addressed or spoken about during the participants' storytelling.

Workshop three: Dolphin Research Institute conversations and tour

The third workshop was held at the DRI main office on the edge of the mangroves of Western Port Bay, where the participants were asked:

What was one thing that you took away from our last workshop in regard to actioning or implementing ESE into your classrooms?
What information or support do you need to help you action your plans?

Once again, dedicated time for conversational storytelling was provided at the beginning of the workshop, with more emotional stories of despair about a lack of ESE in the classroom becoming more visible amongst the participants. A pragmatic conversation that acknowledged the barriers to early career teachers started the DRI presentation, and highlighted the possibility of DRI programmes as examples of how ESE could be enacted. The discussion was followed by a walking tour of the mangroves which, once again, provided more space and time for reflective storytelling on what was *not* happening in teaching ESE in the participants' educational settings. As another refined intervention, the back of the agenda for workshop three provided space for a highly modified action plan asking for 'one practical or simple strategy or resource or activity that could be implemented in your classroom' that could then be shared with the other participants (see Figure 5.3). During this workshop, the researchers were careful not to put any pressure on the participants to contribute any ideas for this template. No participant filled out this modified action plan nor suggested they would do so in the coming weeks.

Workshop four: CERES conversations

Workshop four was held again at CERES and provided another opportunity for emerging shifts in participants' thinking about ESE and further questioning of their role in its enactment in their particular context. The CERES representative provided more advice and valuable resources, including how to present an economic case for sustainability to the school principal or centre director, using a *Taking effective action* model. Various 'next steps' were tentatively suggested to the participants during this workshop, such as a group presentation of ESE ideas to their school principals and centre directors to showcase what might emerge from the project, and/or the possibility of presenting the barriers and challenges early career teachers face to a Community Partnership ThinkTank that was to be arranged for later in the year. However, each of these ideas was quickly and definitively rejected by all of the participants (see Figure 5.3). As a clear shift in direction for the research interventions, the researchers were beginning to understand that the uncompleted action plan templates were a symbolic manifestation of the participants' sense of powerlessness to enact ESE in their educational settings.

Workshop five: DRI conversations and tour

The fifth and final community of practice workshop was held at a primary school where the DRI ambassador programme *i see, i care* was a central feature of their whole-school approach to sustainability. Practical ideas on how to write simple songs with children using climate change messages and a tour through the school's award-winning kitchen garden was the focus of this workshop. The participants were appreciative of the opportunity to witness the kitchen garden programme as well as their visible whole-school approach to sustainable practices. There were three areas of refinement on the research interventions that occurred in this workshop. First, the online blog was dismantled due to a lack of use and interest; second, the decision was made that no agenda would be written for the final workshop, nor was any action plan requested, mentioned or offered; and third, participants' reflections on the overall project were requested for a reflective evaluation as part of the third phase of the research design (see Figure 5.3). Only three of the 11 participants sent through a reflective evaluation of the project to the researchers at a later date.

Workshop six: ESE community partnerships ThinkTank of ideas

Workshop six was designed to be a Community Partnership ThinkTank of potential ESE ideas, as depicted on the research design model in Figure 5.2. We had initially thought the research participants would be interested in presenting their ESE plans, ideas or work to an audience of principals and centre directors to showcase their ESE work during this forum. However, in reality the participants had rejected this idea outright, and so a further refinement of the research interventions was needed (see Figure 5.3). As a result, the ThinkTank workshop did not involve the research participants at all, but rather provided an opportunity for an extended group of ESE community partners (for example, BirdLife Australia, Melbourne Zoo, Cool Australia) to come together and discuss what they considered important supports and resources that could assist in the enactment of ESE in educational settings.

However, what became evident during some of the ThinkTank discussion groups on the day was a lack of expectations on what an early career teacher could do in regard to ESE during their first few years as they 'settled into the teaching profession' (workshop six ThinkTank). It was at this late stage in the research that we as researchers finally realised that this powerful assumption was what the participants were trying hard to push back against throughout the course of the whole research project and throughout their first year of teaching that we had just witnessed. Core to this epiphany was the realisation of how deeply embedded these power structures are in the subconscious of all those involved in educational systems generally.

Summary

Originally, we had envisaged the participants would be encouraged to use the information, advice and resources garnered from the workshops to work on ESE projects in their educational settings. The persistent provision of various templates during the first three workshops seeking populated action plans from the participants was testament to this intention. However, it became increasingly apparent over the course of the study that these action plans would not eventuate, not at this time and in this context. It is interesting to note how the language of the researchers' requests and interventions changed over time; from 'email plans asap' to 'how are you going?' and then, 'what do you need?' to finally an acknowledgement that 'not completing any plan' was a notable and important aspect of the study. Each iterative intervention represented these changes in the way the workshops were co-constructed, with increasingly less focus on the action plans and more on the conversational storytelling in an emotionally 'safe' place. The close examination of emerging teacher identities at a time when they were clearly feeling fragile, vulnerable and challenged by their experiences can

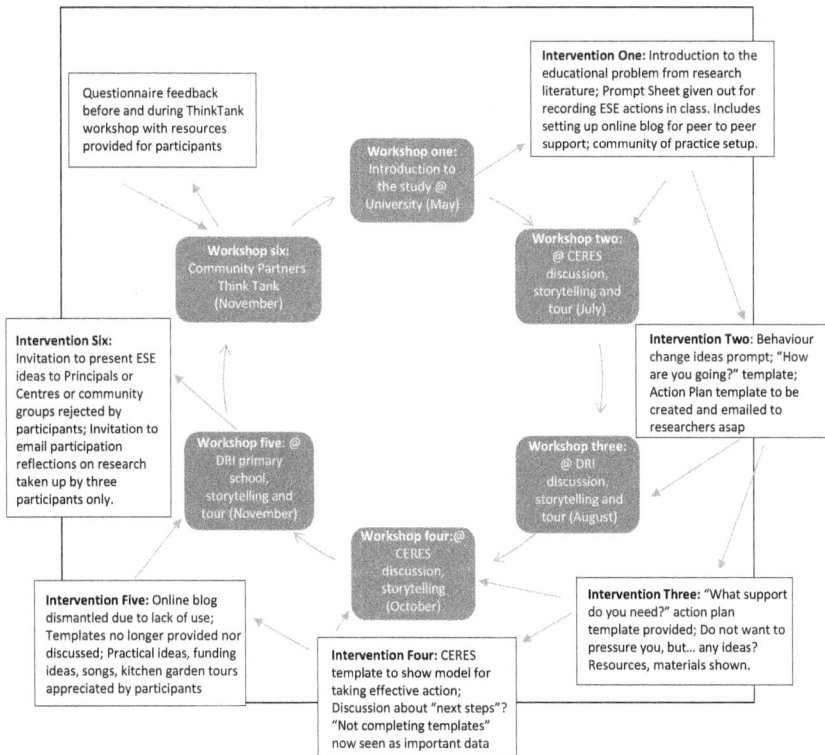

FIGURE 5.3 Visual representation of the research by design model for our study running parallel with the interventions that were implemented and refined throughout the iterative workshops. (Visual representations by Barnes, Moore & Almeida)

be seen to be a 'sensitive topic' that demands careful attention by researchers. This is especially true when the focus of the research is one that early career teachers have held as a critically important element of their identity—personally, professionally, and environmentally—and then to realise they felt incapable of making any 'real' progress in this area in their own educational setting.

Note

1 Please note that permission has been sought and granted for these two community organisations to be named rather than use a pseudonym.

References

Bamberg, M., & Andrews, M. (2004). *Considering counter narratives: Narrating, resisting and making sense*. Amsterdam, NLD: John Benjamins Publishers.

Bell, J. S. (2002). Narrative inquiry: More than just telling stories. *TESOL Quarterly, 36*(2), 207–213.

Bhabha, H. K. (1994). *The location of culture*. Abingdon, UK: Routledge.

Centre for Education and Research in Environmental Strategies (CERES). Retrieved from https://ceres.org.au/about/

Clandinin, D. J. (2013). *Engaging in narrative inquiry*. Walnut Creek, CA: Left Coast Press.

Clandinin, D. J., & Connelly, J. (2000). *Narrative inquiry: Experience and story in qualitative research*. San Francisco, CA: Jossey-Bass Inc.

Clandinin, D. J., & Rosiek, J. (2007). Mapping a landscape of narrative inquiry: Border-land space and tensions. In D. J. Clandinin (Ed.), *Handbook of narrative inquiry: Mapping a methodology* (pp. 35–75). Thousand Oaks, CA: Sage publications.

Cohen, L., Manion, L., & Morrison, K. (2011). *Research methods in education* (7th ed.). Abingdon, UK: Routledge.

Connelly, M. F., & Clandinin, J. D. 2006. Narrative inquiry. In J. L. Green, G. Camilli & P. B. Elmore (Eds.), *Handbook of complimentary methods in education research* (pp. 477–487). Mahwah, NJ: Lawrence Erlbaum Associates Inc.

Cotton, W., Lockyer, L., & Brickell, G. J. (2009). A journey through a design-based research project. In G. Siemens & C. Fulford (Eds.), *Proceedings of World Conference on educational multimedia, hypermedia and telecommunications 2009* (pp. 1364–1371). Chesapeake, VA: Association for the Advancement of Computing in Education.

Creswell, J. W. (2007). *Qualitative inquiry and research design: Choosing among five approaches* (2nd ed.) Thousand Oaks, CA: Sage Publications.

Davis, J., & Elliott, S. (2009). Exploring the resistance: An Australian perspective on educating for sustainability in early childhood. *International Journal of Early Childhood, 41*(2), 65–77.

Denzin, N. K., & Lincoln, Y. S. (2008). Introduction- the discipline and practice of qualitative research. In N. K. Denzin & Y. S. Lincoln (Eds.), *Collecting and interpreting qualitative materials*. (3rd ed., pp. 1–44). Thousand Oaks, CA: Sage publications.

Dolphin Research Institute (DRI). Retrieved from https://www.dolphinresearch.org.au/dri/

Golombek, P. R., & Johnson, K. E. 2004. Narrative inquiry as a mediational space: Examining emotional and cognitive dissonance in second-language teachers' development. *Teachers and Teaching: Theory and Practice, 10*(3), 307–327.

Hollingsworth, S., & Dybdahl, M. (2007). Talking to learn: The critical role of conversation in narrative inquiry. In J. D. Clandinin (Ed.), *Handbook of narrative inquiry: Mapping a methodology* (pp. 146–176). Thousand Oaks, CA: Sage Publications.

Jørgensen, R., Edwards, K., & Ipsen, C. (2018). Intentional development of communities of practice: Improving knowledge sharing and work guidelines. *European Conference on Knowledge Management.* Implementation and performance management, DTU Management. Academic Conference & Publishing International Ltd., Denmark.

Kramp, M. K. (2004). Exploring life and experience through narrative inquiry. In K. B. de Marrais & S. D. Lapan (Eds.), *Foundations of research: Methods of inquiry in education and social sciences* (pp. 103–123). Mahwah, NJ: Lawrence Erlbaum Assoc. Inc.

Kuntz, A. M., & Presnall, M. M. (2012). Wandering the tactical: From interview to intraview. *Qualitative Inquiry, xx*(x), 1–13.

Nolan, A., Macfarlane, K., & Cartmel, J. (2013). *Research in early childhood.* London, UK: Sage Publication.

Plomp, T., (2013). Chapter 1: Educational design research: An introduction. In T. Plomp & N. Nieveen (Eds.), *Educational design research: Part A: An introduction* (pp. 10–51). Enschede, The Netherlands: Netherlands Institute for Curriculum Development.

Reeves, T. (2006). Design research from a technology perspective. In J. Van den Akker, K. Gravemeijer, S. McKenney & N. Nieveen (Eds.), *Educational design research* (pp. 52–66). New York, NY: Routledge.

Riessman, C. K. (2008). *Narrative methods for the human sciences.* Thousand Oaks, CA: Sage Publication.

Sherwood, S. A. S., & Reifel, S. (2010). The multiple meanings of play: Exploring preservice teachers' beliefs about a central element of early childhood education. *Journal of Early Childhood Teacher Education, 31*(4), 322–343.

Somers, M. (1994). The narrative construction of identity: A relational and network approach. *Theory and Society, 23,* 605–649.

Wang, F., & Hannafin, M. J. (2005). Design-based research and technology-enhanced learning environments. *Educational Technology Research and Development, 53*(4), 5–23.

Wenger, E., McDermott, R., & Snyder, W. M. (2002). Cultivating communities of practice: A guide to managing knowledge. Boston, MA: Harvard Business School Press.

6

COMMUNITY PARTNERSHIPS—'JUST SNEAK IT IN'

Subversive ways to include ESE

Box 6.1 Extract from community partner in workshop two

I think that's where we can be much cleverer and dare I say... subversive. Leadership hasn't come and said, 'You must not teach sustainability to these kids', they just want them to bring up their reading and literacy [skills]. Well, do it whatever way you can, and if you can just sneak it in, then all the better...

(Community partner, workshop two)

Introduction

This chapter begins with a purposefully chosen provocation, a challenging quote from a community partner. This quote provokes an emotional response which prepares us to begin to understand the personal and professional Environmental and Sustainability Education (ESE) experiences revealed through the early career teachers' stories and the community partner responses.

Community partnerships have played an important role in promoting ESE within the community (Allen-Gill, Walker, Thomas, Shevory & Elan, 2005; DePetris & Eames, 2017; Kawabe et al., 2013; Moore, O'Leary, Sinnott & O'Connor, 2019). There has been a strong tradition of educators, from schools, early childhood centres and universities, partnering with community partners to create meaningful place-based learning opportunities for their learners (Allen-Gill et al., 2005; Chan, Matthews & Li, 2018; DePetris & Eames, 2017; Moore et al., 2019). This chapter examines the knowledge and advice elicited by the key community partners in this study, which not only provides important professional and ecological suggestions but highlights the value of community partnerships in ESE.

Shared vision and ideas: creating meaningful community partnerships

A number of studies have argued that community partnerships, which have been established to promote ESE within their communities, allow for the local community to come together in a meaningful way to share a common vision and purpose (Allen-Gill et al., 2005; DePetris & Eames, 2017: Kawabe et al., 2013; Moore et al., 2019) and therefore strengthen community spirit. Allen-Gill et al. (2005) argued that in their US study that brought together researchers at a university and residents of an 'EcoVillage,' a model sustainable community located in close proximity to their university campus provided the opportunity to create a 'critical mass' that had the capacity to generate significant change (p. 401). This partnership opened the door for several key and mediating individuals to begin building a mass of people who shared a similar drive to work towards meaningful ecological change in their community. As Best (2019) argues, 'the most important resources in a local community is its people' (p. 23), and it is identifying and then harnessing the capacities of individuals that allows for a community to grow stronger. Best (2019) acknowledges that there are community members who are positioned to be connectors within the community:

> …the connectors are the people with the vision to see both potential and positives and also in a position to mobilise these assets through their connections…. it is an inherently social approach, which places great faith and emphasis on individuals and their ability and willingness to commit to actions and activities for the benefit of their neighbourhood and what they perceive to be their communities.
>
> *(p. 26)*

In line with Best (2019), several studies acknowledge that the learners are key connectors in community change (DePetris & Eames, 2017; Kawabe et al., 2013; Reeves, 2019). In New Zealand, DePetris and Eames (2017) explored the perspectives of stakeholder representatives from five educational organisations and four community organisations involved in an 18-month project called *Kids greening Taupo*. While the stakeholders agreed that the school-aged learners who took part in this project were key connectors and/or mediators of change, they also identified the role that the stakeholders play in creating authentic learning materials. Additionally, several participants acknowledged the key role of collaborative planning across the participating organisations but felt that there needed to be 'more "quick wins" so that the children and young people did not get dragged down by the red tape of bureaucratic planning and the long-term timeframe necessary for genuine restoration outcomes to be fully recognised' (p. 182). While learners were seen as keys to change, they recognised the need for learners to view ecological change in the short term to keep them engaged. Therefore, learners (and teachers) needed to be able to see the value of small wins and that small steps would get them closer to significant change.

Set within a UK university setting, Reeves (2019) designed a pilot programme that provided open workshops to both university learners in a UK university and the local community. The content focus was on sustainability action and the aim of the project was to facilitate knowledge exchange, social networking and reflective practices among this diverse group. While learners played a mediating role, all participants, including the local community members, found that this participatory approach was valuable, particularly in how the diversity of the group allowed opportunities to link theory with practice.

In addition to learners being positioned as agents of change within the community, the literature suggests that educators play a key role in providing opportunities for learners to engage with ESE concepts and ideas. Several studies highlight the need for continued professional development among educators to increase confidence, knowledge and capability to embed ESE within the curriculum (Chan et al., 2018; DePetris & Eames, 2017; Moore et al., 2019; Pedersen, 2017).

One approach that has been used to share expertise among educators has been through the creation of a Community of Practice (CoP), allowing for informal learning opportunities within the area of ESE (Moore et al., 2019; Pedersen, 2017). A CoP consists of 'groups of people who share a concern, a set of problems, or a passion about a topic, and who deepen their knowledge and expertise in this area by interacting on an ongoing basis' (Wenger, McDermott & Snyder, 2002, p. 4). Therefore, an ESE CoP allows for shared learning in both the concepts and practices of ESE. Moore et al. (2019) view CoPs as key for 'knowledge development, research, professional development and the sharing of knowledge and experience' allowing 'school staff to harness the expertise and skills they require to ambitiously work on the practice of sustainable schools' (p. 1759).

In the context of their study, they suggested that education partnerships were vital as universities had the resources and capabilities to support the professional development and training of educators in a range of education settings. They argue that there is a need for the distribution of ideas, expertise and resources to further ESE initiatives, particularly in supporting schools and centres to embed ESE concepts within the curriculum.

Place-based learning for ESE

There is a strong tradition, particularly in the context of early childhood education that argues the important role of *place* in the development of children's identities (Pettifer, 2019; Shannon & Galle, 2017; Smith & Sobel, 2010; Sobel, 2014). Place-based learning embraces the role of *place*, particularly when viewing place as one's local community and environment, as a valuable and authentic space for learning. Shannon and Galle (2017) suggest that place-based learning is a response to the abstractions of school which has attempted to address the 'alienating effects of non-permeable classroom walls that separated students from their communities rather than embedding them within the ecologies where they are located' (p. 5). When considering sustainability, place-based learning provides opportunities for learners to situate ecological problems and solutions within their communities. Pettifer (2019) argues that a connection with nature 'is recognised as a motivator for environmentally preferred behaviours and to develop active environmental citizens' (p. 18), and that a learner's sense of connection and/or love of place are critical foundations for developing ESE values.

The creation of a number of community partnerships has been the result of educators, both school or early childhood teachers and/or university lecturers, attempting to provide meaningful place-based learning opportunities for their learners (Allen-Gill et al., 2005; DePetris & Eames, 2017; Chan et al., 2018; Moore et al., 2019). In the contexts of both New Zealand (DePetris & Eames, 2017 and China (Chan et al., 2018), organisations have come together to provide learners with authentic learning opportunities—more specifically, using nature reserves as places of learning. Chan et al. (2018) found that ESE, focusing on concepts relating to biodiversity and agriculture, took on new meanings for local school children when discussed in relationship to the nearby Caohai Nature Reserve. Given that the livelihood of 20,000 farmers depends on this reserve, a local community partnership between local farmers, non-government organisations, nature reserve staff and local schools has led to opportunities to embed ESE learning into the curriculum of local schools:

> As the local community accepted responsibility and took leadership of the program, an empowered atmosphere developed, which has provided a foundation for continued focus on resource conservation and the development of EE [Environmental Education] programs in local schools.
>
> *(Chan et al., 2018, p. 178)*

Similarly, DePetris and Eames (2017) argue that community programmes can have a meaningful impact on the local community, but these grassroots efforts require partnerships that extend beyond these immediate communities:

> ...programs like Kids Greening Taupo can help improve community and environmental welfare through the distribution of resources and expertise from regional and national organisations to grass-roots efforts. In this way, children and young people are afforded authentic learning opportunities that simultaneously can enable them to contribute to making a positive difference today.
>
> *(p. 185)*

Therefore, the use of place-based learning in education allows teachers to embed key ESE concepts within the curriculum while connecting learners' learning with their local community, making a powerful connection between place, ESE and identity. However, in order to promote grassroots initiatives that provide the opportunities to connect the local community with ESE learning and build environmental identities, key partnerships are required to help in distributing expertise and resources.

Our research project: building relationships with community partners

An integral element of the overall design and set-up of the research project this book is based on was the inclusion of representatives from two key local ESE community partnership organisations into the CoP workshops. As previously explained in Chapter 5, both the Centre for Education and Research in Environmental Strategies (CERES) and the Dolphin Research Institute (DRI) as key sustainability-orientated organisations had been engaged in the past with the authors through the university's Sustainability unit. For example, DRI provided experiential excursions for pre-service teachers to local Victorian beaches and foreshores, leading discussions on the impact of climate change on the local environment, while CERES provided resources and advice on ESE initiatives for pre-service teachers.

For this research project, one community partner representative from DRI and another one from CERES committed to attending and/or hosting each of the workshops to be part of the CoP conversations. In each workshop, at either Hastings or Brunswick, the community partners took turns to present more focused, contextually meaningful information and advice at their own organisation's site. In line with the research by design methodology, the plan was for the nominated community partner to provide iterative and inspirational advice and tips over the course of the project to prompt and ignite participants' ideas, questions and inquiries, which, we had assumed, would lead to the co-construction of the teachers' own projects to enact ESE in their classrooms.

As a research relationship started to emerge, develop and build between all of the project participants (that is, 11 early career teachers, three researchers and two community partners), the significantly clear role the community partners played in each iterative workshop conversation was especially noteworthy. Building on the commentary and advice given in the previous workshop, each community partner designed and organised relevant content, resource materials and tours to demonstrate to the early career teachers what was available in practical terms for teachers and teaching in the field of ESE. Upon analysis of the storytelling at the beginning of each workshop, both the community partners appeared to listen intently to the teachers' stories; contributed in respectful ways; and made insightful prompts to extend the conversations rather than speaking over the teachers or simply telling them 'what to do.' For example, one of the community partners responded to a teacher's poignant story about a large gumtree being removed from his school playground, asking, 'How did that make you feel?' While in another workshop, a teacher was tearfully talking of her despair that her curriculum was completely NAPLAN driven and had no space for ESE. On this occasion, the community partner was highly responsive and asked respectful questions to prompt everyone's thinking about how to shift the school principal and other teachers' thinking about this situation.

Over time, this respectful attitude permeated the culture of the group as a safe 'third place' (Bhabha, 1994) and created a strong sense of trust and collegiality amongst all the participants within the CoP. From the teachers' positive comments, as indicated in Chapter 5 and in the example given below, this storytelling time at the beginning of each workshop unexpectedly became the most important aspect of the whole project for the teachers:

> …it is great having these conversations because I always leave feeling energised after having a chat with like-minded people… it's like a dopamine boost, makes you feel good coming here knowing that there's people out there that think the way you do and are passionate about this…
>
> *(Waruni, workshop three)*

In any narrative inquiry, trust is a critical component of the 'reciprocal relationship' developing amongst the research co-constructors (Clandinin & Connelly, 2000), where it is hoped participants would feel safe to talk about their lived experiences during interactive, responsive conversations. In this narrative inquiry, a trusting research relationship was also seen to be a critically important aspect of the CoP workshops. This was because it was through the teachers' heartfelt stories that the deeper meanings of their experiences were articulated and made visible for their own reflective understandings and for the analysis of the storied data.

The teachers' and the community partners' stories: a three-dimensional analysis

Through the lens of this community partnership, three distinct themes became evident in the three-dimensional narrative inquiry analysis of the teachers' stories and the CoP conversations at this time and place. The first temporal theme focused on the need for teachers to take *baby steps*, with the understanding that any shift in thinking towards ESE in schools and early childhood settings is most commonly a *long, slow journey* over time (Community partner, workshop two). The second theme is based on societal attitudes to ESE, suggesting the teachers need to push back against negativity by *play[ing] the game* (Community partner, workshop two) using social marketing and behaviour change techniques to *market* the benefits of ESE to time-poor teachers and NAPLAN-focused principals. Finally, the third theme centres on different place-based contexts with the learners *holding the key* in the success or constraint to embedding ESE into centres or schools. Remembering that in line with a narrative inquiry of three-dimensional analysis, each dimension is analysed simultaneously with inevitable crossovers between each of the dimensions.

'Always take baby steps': the slow journey towards ESE

Within the CoP space, the overarching theme was that the teachers needed to *slow down* and take *baby steps* (Community partner, workshop two) rather than thinking they could make significant change towards ESE in their schools, classrooms or early childhood centres in a hurry. Sadly, one of the early career teachers revealed that she wished she had known about the need for *baby steps* earlier, as she had rushed in with an ESE idea for the whole school and was reprimanded by the principal for not asking for approval beforehand. In a reflective response to this teacher's disheartening experience, the community partner suggested:

> You will always get some dinosaurs and it's brutal... you've got to be aware that some teachers have never worked outside of a school or academic environment and so they have a very insular view of those sorts of things, and things can get incredibly petty... So, my advice would be that you **always take baby steps** and you find out where the hierarchies are and you find out what you need to do to get approval before you jump. And I know that's after the horse has bolted...
>
> *(Community partner, workshop two)*

Similar to De Petris and Eames' (2017) study, where the teachers needed to see that significant change was still possible following 'small steps' (p. 182), this notion of 'baby steps' seemed to disrupt many of the early career teachers' sense of urgency around sustainability bolstered by their strong environmental identities. In later workshops, the use of 'baby steps' became a common phrase as part of the shared language and understanding amongst the CoP. This was especially noticeable when one community partner led the group in an interactive discussion about how to work through the iterative six (baby) steps of the CERES Model for Taking Effective Action as seen below:

1. '**Start a buzz**' about a sustainability issue by 'starting slowly through baby steps' in a casual way to 'plant the seed';
2. '**Offer hope**' by reassuring others that it is possible to find a solution together, set up a small group or portfolio by recruiting others who are interested;
3. '**Find the bright spots**' by looking at where the school/centre is already successfully engaged in sustainable practices that could be replicated;
4. '**Designing a sticky solution**' that must be easy (and sustainable), measurable, with clear directions, has in-built social pressure and an element of

fun; find out what excites the learners (and others) so they are more likely to want to be involved;

5. '**Find the right inviters**' by finding the 'clincher deal' so that you are able to 'immobilise others' possibly through the 'children and their parents'; this is the stage where you can '**make the announcement and officially start**' a project;

6. '**Expand the comfort zone**' by holding a celebration of success, for even the 'small[est] wins' (De Petris & Eames, 2017, p. 182); building ownership and gathering feedback, adding more actions and reflecting on what has been achieved.

The concept of 'baby steps' aligns well with Bourdieu's concepts of habitus and field, suggesting that change, a product of agency, occurs gradually as the habitus (one's way of doing and thinking) and field (the social context) interact and shape one another. Bourdieu (1977) argues that while habitus is shaped by the field, it can also *shape* the field. Therefore, teachers can introduce ESE concepts to the field, and these concepts might slowly and gradually be introduced and accepted. Through these small steps, teachers enact their agency and reshape their identities by slowly introducing new ways of thinking and doing that ultimately can have a powerful influence over the field.

Following on into the next workshop, and in alignment with Reeves' (2019) participatory workshops where knowledge is exchanged, reflective practices are enabled and social networking encouraged, there was a palpable increase in understanding amongst the teachers on the processes of 'baby steps' along a continuum over a 'long slow journey.' This was evident in the participants' nods and reciprocal smiles, whilst ideas and stories flowed back and forth around the room about how to move slowly towards embedding ESE into their classrooms and what that may look like over time. At this point, one community partner astutely suggested:

> …slow down and suddenly, things are there [that you hadn't noticed before]…I don't teach sustainability, it is just there…Start small, as (the other community partner) says, even if it is just in your own classroom, start to create the buzz…
>
> *(Community partner, workshop three)*

While the mood had definitely lifted with the knowledge that they could slow down and not rush into ESE actions, the teachers still expressed their shared frustration in not enacting ESE during their first year of teaching, as they had assumed they would do in line with their personal environmental identity.

'Play the game'—the subversive inclusion of ESE

What was becoming increasing apparent throughout the project was how much the participants looked forward to the CoP conversations. This was particularly so when the original plan to enact their own ESE project gradually dissipated as the researchers shifted their thinking, adapted the methodological interventions and purposefully changed the language about the focus of the study. Instead, a cohesive support group formed and enclosed around the participants, where previously hidden and silenced stories were listened to and welcomed. Craig and Huber (2007) suggest one of the key advantages of a narrative inquiry is where an in-depth, reciprocal research relationship is seen as valuable rather than an objective researcher/bystander approach; whilst other narrative inquiry researchers suggest the research relationship based on co-construction helps to 'capture different... often otherwise hidden elements of stories' (Connelly & Clandinin, 2006, p. 483).

In creating a safe place for silenced stories, a turning point for the teachers occurred when one community partner strongly advocated for the teachers to 'play the game' and to become much more 'subversive' in their attempts to include ESE into their overcrowded curriculum, saying:

> I think that's where we can be much cleverer and dare I say... subversive. Leadership hasn't come and said, "You must not teach sustainability to these kids", they just want them to bring up their reading and literacy [skills]. Well, do it whatever way you can, and if you can just sneak it in, then all the better...
>
> *(community partner, workshop three)*

Once again, insightful comments from the community partners seemed to disrupt the teachers' metacognitive thinking about their teaching in a way they were not expecting. While laughter erupted around the notion of subversive and 'sneaky' sustainable practices, it was also cause for serious contemplation and a reimagined sense of agency within their classrooms. Simple ideas around including small, hidden sustainability segments, ideas and/or routines in each teacher's classroom were discussed, contemplated and considered as possibilities. These simple ideas were akin to small streams (even trickles) that all add up to a stronger flow to the larger confluence of identities. The conversation continued:

> And you learn to play the game a little bit because ultimately what you want is the objective [to embed ESE into the curriculum]... How you go about it doesn't really matter as long as you get what you want in the end...
>
> *(Community partner, workshop three)*

Bourdieu refers to the field as being a game—one that is governed by rules (1977). He argues that these rules can be explicit and implicit and the players must endeavour to understand how best to play the game in order to win. Bourdieu and Wacquant (1992) suggest that there are strategies for playing the game. These hidden, and somewhat subversive, tactics may be a way to play the rules of the game yet all the while embedding ESE into the curriculum. Specific strategies mentioned were social marketing and behaviour change techniques to enable sustainability to be 'hidden' within their day-to-day teaching practices also became a common thread in the community partners' conversations, where they suggested:

> That's where social marketing comes in… even if you can't do anything tangible, you're still a role model, bring your own practices into the classroom
>
> *(Community partner, workshop three)*

Researchers and writers in behaviour change techniques claim 'marketing' has the potential to make a positive contribution to changing collective behaviour around sustainability (Gentile, 2010). Gentile (2010) argued that 'psychological tricks help move people from one step to the next' when advocating for major 'health' behavioural changes; and, that we should not be 'afraid to use emotion as a motivator' for the need for change (p. 97).

This approach was clearly argued for in the community partners' conversations in the latter workshops, such as when one of the community partners suggested using 'contrasting principles' (Gentile, 2010, p. 97) by comparing local schools to the principals, for example, talking about the '*great energy saving work another school is doing just down the road*'! (community partner, workshop four). Social marketing techniques are said to work by first gaining people's 'emotional response' as the 'brain remembers things that are emotional' and then 'building an attitude' towards the motivation to change their behaviour (Gentile, 2010, p. 97).

Both community partners advocated for the use of emotion and other social marketing techniques by suggesting teachers focus on practices that are applicable to an educational context. For example, one community partner suggested that the teachers '*find sneaky tools*' to integrate sustainability into literacy and numeracy lessons, such as through mathematical calculations on energy use and/or reading about the local biodiversity. In this way, the principal's push for numeracy and literacy is still addressed, but with the seamless inclusion of sustainability, not as a separate, disconnected and/or neglected subject. In other words, teachers play the rules of the game while also getting to play it the way they would like.

'*The kids are the key*'– the influence of learners on ESE in the classroom

References to learners became a common narrative thread throughout the CoP workshops. First, because the teachers and community partners recognised learners as the ones who will be faced with the environmental crisis in the future, as seen in the following quote:

> …it has to become apparent to people that it's not just about a bunch of environmental tree huggers and hippies anymore, this is going to happen

> in the next 20, 30, 40 years to this planet that we are living on, that these kids are going to inherit these problems…
>
> *(Community partner, workshop two)*

And second, because both the community partners strongly believed the learners could convince the school and centre leaders of the need to focus on sustainability in their educational settings, evident here in the following quotes:

> The kids are the key…they are great at pressuring adults…the kids are a really good key to getting progress because, it's not the teachers' futures, it's the kids' future, and so they have to take ownership of it on your behalf.
>
> *(Community partner, workshop two)*

And again in the following workshop:

> The kids are your greatest allies to be honest. They can run it, through peer pressure, saying to the Principal, 'we want to talk about this'…who will inevitably respond by saying… 'sure you can'…
>
> *(Community partner, workshop three)*

The notion that learners are key connectors in promoting ESE aligns with a number of other studies (DePetris & Eames, 2017; Kawabe et al., 2013; Reeves, 2019); however, this contrasts with the views of the Prime Minister of Australia at the time, Mr. Scott Morrison. Mr Morrison declared that children should be learning in the classroom not holding placards in Climate Change Strike demonstrations (as discussed in Chapter 1). However, the community partners strongly disagreed. Instead, they were convinced the 'kids' in each respective classroom (or strike) were the 'key' in convincing adults that ESE should be an essential part of their learning.

Once again using the principles of social marketing through provoking an emotional, uncomfortable response from adults, photos of learners (in clearly recognisable school uniforms) pleading for the future of the planet has the potential for creating changes to attitude and behaviour in the school and early childhood leaders.

Looking to the future: hopeful applications

Positive messages and practical, respectful advice from the community partners within the safety net of the CoP workshops were a surprisingly critical element of the study. Each of the participants noted how grateful they were for the conversations and the advice that was given by each of the community partners, despite the early career teachers' lack of ESE action undertaken. During the workshops, the early career teachers felt validated, listened to and respected for their personal, professional and ecological identities rather than feeling they were 'incapable of any action in their first year of teaching,' as the dominant discourse around early career teachers suggests (community leader, ThinkTank workshop).

In a final message in the last workshop, both community partners declared that the way to encourage the implementation of ESE in educational settings was to *give teachers a* [positive] *reason* to engage with sustainability, and that the ideas needed to be *ironically sustainable… and not too hard or expensive* to be able to keep going rather than giving up at the first hurdle. One community partner suggested that if sustainable practices cost money:

> …it becomes a luxury unfortunately (*if it costs any money*)… positivity is really important, you have to give them positive reasons why this is a good thing… **benefit, benefit, benefit**, what's in it for them, what's in it for the school, why are they going to benefit and you're much more likely to get them saying, 'Oh really, I hadn't even thought of that benefit, that's a really good one'…
>
> *(Community partner, workshop six)*

Similarly, the other community partner confirmed this line of positive thinking by suggesting the new career teachers could bring the conversation with the school and centre leaders back to how much money can be saved through being aware of water leaks, energy overuse and natural play-spaces rather than expensive, plastic-dominated playgrounds. This chapter suggests that teachers have the capacity and the opportunity to enact agency through small steps that may gradually shape how their schools and centres view ESE and its place within the curriculum. While they must play the game, they can utilise strategies, such as being a role model introducing changes slowly, and mobilise learner support for change.

References

Allen-Gill, S., Walker, L., Thomas, G., Shevory, T., & Elan, S. (2005). Forming a community partnership to enhance education in sustainability. *International Journal of Sustainability in Higher Education, 6*(4), 392–402.

Best, D. (2019). Australian origins: building bridges and community connections. In D. Best (Ed.), *Pathways to recovery and desistance: The role of the social contagion of hope* (pp. 23–47). Bristol: Bristol University Press.

Bhabha, H. K. (1994). *The location of culture*. Abingdon, UK: Routledge.

Bourdieu, P. (1977). *Outline of a theory of practice*. Cambridge: Cambridge University Press.

Bourdieu, P., & Wacquant, L., (1992). *An invitation to reflexive sociology*. Chicago, IL: University of Chicago Press.

Chan, Y-W., Mathews, N., & Li, F. (2018). Environmental education in nature reserve areas in southwestern China. *Applied Environmental Education & Communication, 17*(2), 174–185. doi:10.1080/1533015X.2017.1388198

Clandinin, D. J., & Connelly, J. (2000). *Narrative inquiry: Experience and story in qualitative research*. San Francisco, CA: Jossey-Bass Inc.

Connelly, M. F., & Clandinin, J. D. (2006). Narrative inquiry. In J. L. Green, G. Camilli & P. B. Elmore (Eds.), *Handbook of complimentary methods in education research* (pp. 477–487). Mahwah, NJ: Lawrence Erlbaum Associates Inc.

Craig, C. J., & Huber, J. (2007). Relational reverberations. In D. J. Clandinin (Ed.), *Handbook of narrative inquiry: Mapping a methodology* (pp. 251–279). Thousand Oaks, CA: Sage Publications.

DePetris, T., & Eames, C. (2017). A collaborative community education model. *Australian Journal of Environmental Education, 33*(3), 171–188. doi:10.1017/aee.2017.26

Gentile, D. A. (2010). Selling ideas, attitudes, and behaviors. *Journal of Agromedicine, 15*, 96–97.

Kawabe, M., Kohno, H., Ikeda, R., Ishimaru, T., Baba, O., Horimoto, N., Kanda, J., Matsuyam, M., Moteki, M., Oshima, Y., Sasaki, T., & Yap, M. (2013). Developing partnerships with the community for coastal ESD. *International Journal of Sustainability in Higher Education, 14*(2), 122–132. doi:10.1108/14676371311312842

Moore, M., O' Leary, P., Sinnott, D., & O'Connor, J. (2019). Extending communities of practice. *Environment, Development, and Sustainability, 21*, 1745–1762. doi:10.1007/s10668-018-0101-7

Pederson, K. (2017). Supporting collaborative and continuing professional development in education for sustainability through a communities of practice approach. *International Journal of Sustainability in Higher Education, 18*(5), 681–696. doi:10.1108/IJSHE-02-2016-0033

Pettifer, L. (2019). Making sustainability happen. *The Social Educator, 37*(2), 14–28.

Reeves, A. (2019). Classroom collaborations: Enabling sustainability education via student-community co-learning. *International Journal of Sustainability in Higher Education, 20*(8), 1376–1392. doi:10.1108/IJSHE-11-2018-0220

Shannon, D., & Galle, J. (Eds.) (2017). *Interdisciplinary approaches to pedagogy and place-Based Education.* London, UK: Palgrave MacMillan.

Smith, G., & Sobel, D. (2010). *Place- and community-based education in schools.* New York, NY: Routledge.

Sobel, D. (2014). *Place-based education: Connecting classrooms and communities.* Great Barrington, MA: Orion.

Wenger, E., McDermott, R., & Snyder, W. M. (2002). *Cultivating communities of practice: A guide to managing knowledge.* Boston, MA: Harvard Business School Press.

7

THE WHOLE-SCHOOL APPROACH IN ESE SCHOOLS

'I'm the lucky one here'

Box 7.1 Extract from Alison in workshop two

This is really awful… I had an excursion to CERES last year with the Grade 6s and I caught up with a student who hadn't yet paid and the parent responded, "Oh yeah but what are you doing there?" and I said, "We're going to go CERES, that's our inquiry topic and they'll learn a few things about sustainability" and she said, "Oh nah, that's okay, I'm just not really into that kind of [environmental] stuff"…

(Alison, workshop two)

Introduction

In positioning the schools discussed in this chapter, we begin by calling attention to the use of terms such as 'tree hugger' and/or 'greenies,' which have been commonly used with negative connotations in dominant discourses to criticise the actions of people who believe in taking action rather than 'just talk' about the need to save trees or indeed, the planet (1 million women, 2015). In confirmation of the negative use of the term, Merriam-Webster dictionary defines the term 'tree hugger' as follows:

> **Definition of** *tree hugger*
> *disapproving*: someone who is regarded as foolish or annoying because of being too concerned about protecting trees, animals, and other parts of the natural world from pollution and other threats.
>
> *(Merriam-Webster, nd)*

We have begun with this provocation because the two early career teachers (Shaun and Alison) and their respective schools (one inner-city school and one semi-rural school, both in high socio-economic status [SES] areas), where a whole-school approach to Environmental and Sustainability Education (ESE) has been implemented, have anecdotally been classified as 'greenie' schools. However, despite the expectation by the majority of these school communities that sustainability is embedded into the curriculum, there is still an element of the parent community who push back against the label as 'greenies' or apparently even worse, 'environmental activists' (Uren et al., 2019), alluded to in Alison's quote at the beginning of this chapter.

Framing the two early career teachers' stories of their ESE experiences within a whole-school approach, national and international research and literature shows this is the most successful way to embed sustainability into the curriculum. Following this, literature around the intentional use of inquiry-based education aligns well with the broader principles of ESE and provides a pedagogical approach that enables the in-depth enactment of ESE in an educational setting. While a three-dimensional narrative inquiry analysis of Shaun's and Alison's stories reinforces the literature around the success of a whole-school approach, what is unique to this study is the way the early career teachers questioned and re-evaluated their own reflexive involvement in ESE in their schools. This is evident in this chapter, first in the non-linear approach to leadership around encouraging sustainability in the schools; second, the social justice ethos of each of the two schools was clearly linked to the overarching principles of ESE; and third, in relation to place-based dimensions, the culture of the school context determines how conducive it is to the development of new ideas and innovative thinking around the inclusion of ESE into the curriculum. The final section of this chapter illustrates how the teachers' stories provide practical and hopeful ways to move beyond the barriers of policy, context and dominant discourses to work towards the successful enactment of ESE into their classrooms.

A whole-school approach to ESE

International and national whole-school sustainability initiatives have called for a push towards a cultural shift in how ESE is understood and implemented within educational settings (Goldman et al., 2018; Henderson & Tilbury, 2004; Salter, Venville & Longnecker, 2011; Warner & Elser, 2015). Goldman et al. (2018) argue that a whole-school approach to sustainability is 'holistic and participatory' and is transformative in how schools integrate ESE into the curriculum, school operations and their relationships and interactions with the local community (p. 1302). Similarly, Warner and Elser (2015) suggest that 'interconnectedness' is needed as it allows for 'interactions, collaborations, and integrations between diverse and relevant disciplines, ideas, and educational stakeholders in order to teach students that our actions may, and often do, result in unintended consequences' (p. 2). This suggests that while local community partnerships are important for a successful whole-school approach (see Chapter 6), it is the wide range of stakeholders, including school leaders, teachers, learners, parents and the local community, that help in creating a positive school culture surrounding ESE (Henderson & Tilbury, 2004; Koç & Bastas, 2019; Salter et al., 2011). Therefore, a whole-school approach to sustainability requires strong school governance (Henderson & Tilbury, 2004) and a school ethos that embraces ESE (Bosevka & Kriewaldt, 2020). Leadership who not only support ESE but view it as an important part of their school culture are able to instil particular attitudes and values among their learners and teachers (Bosevka & Kriewaldt, 2020). These shared values create an environment where changes in practices for a more sustainable future become a more tangible possibility.

Bosevka and Krieswaldt (2020) found that the support of leadership was particularly important with regard to how ESE was integrated into the curriculum: 'The curriculum is a vehicle through which we maintain culture, it is the primary means for transmitting the school's values and expectations and for putting those beliefs into daily practice' (p. 62). As a result, a school's organisational identity, or institutional habitus, is instrumental in determining how they implement (and/or prioritise) ESE. However, it is important to understand that a school's institutional habitus is not shaped in isolation but is shaped in relation to its stakeholders (Bartlett, McDonald & Pini, 2015; Brickson, 2007). Therefore, a successful whole-school ESE approach is reliant on more than just leadership—but the wider community's support. In schools that have a strong ESE ethos, they exert and exercise power through normative and discursive practices, such as through whole-school approaches to curriculum design and implementation as well as professional development (Henderson & Tilbury, 2004).

Barnes, Moore and Almeida (2018) found that even though opportunities for whole-school initiatives were offered to schools in Australia, it required the support from school leadership to prioritise and guide teachers in using the tools and resources available. In contrast, Warner and Elser (2015) argue that in the US, many school leaders of K-12 schools acknowledged the importance of ESE but lacked the tools and/or resources to guide them in developing sustainability programmes.

One way in which schools have initiated and begun the process of making ESE part of their school culture has been by becoming accredited through sustainability certification or award programmes provided to schools which offer frameworks for implementing ESE (see Griffin & Chan, 2015; Henderson & Tilbury, 2004; Metzger, 2015). Examples of national and global certification and award programmes include Australian Sustainable Schools Initiative (AuSSI) (2010), Eco-schools international programme, Israel's Green School Certification (Goldman et al., 2018) and New Zealand Enviroschools and Sustainable Schools in the UK (Sustainability and Environmental Education SEEd, 2020). The push for 'green schools' is not new to education as the early 1900s were shaped by the 'Nature Study movement' as well as 'the Conservation Education movement' (Metzger, 2015, pp. 1–2). However, as certification and award programmes have been created to help guide schools in making environmentally responsible decisions and prompting action, they also provide a framework in order to measure how well schools are meeting ESE outcomes, including reducing energy consumption, integrating ESE concepts into the curriculum and creating local community partnerships (Metzger, 2015).

With a number of 'green school' initiatives and a variety of certification levels (from basic to more advanced), this calls into question how effective these certification programmes are in promoting meaningful change in the classroom. Goldman et al.'s (2018) study examined the environmental literacy of 403 grade six students from eight primary schools in Israel. These schools represented the following four types of schools (two schools were chosen for each category):

1.   Certified as an *ongoing-green* school (the highest ESE certification, certified for 8–10 years);
2.   Certified as a basic *green* school (certified for 6–8 years);
3.   In the process of certification;
4.   Not part of the green school certification programme.

The results of their study found that learners in *ongoing-green* schools had more advanced environmental literacy than their *green* school peers. While the green school learners had been part of this basic *green* programme for the entirety of their schooling, they still demonstrated 'lower-than-expected' environmental literacy. This suggests that the schools which have acted to seek the higher accreditation are most likely committed to ESE in a way that is integral to their school identity and ethos. This school culture provides opportunities for interconnectedness and a shared ESE vision that transforms a school's 'infrastructure, management, curriculum and community outreach' (Goldman et al., 2018, p. 1308) that allows for meaningful learning. Ultimately, this study suggests that the relationship between a school's culture and ethos is strongly linked to the learners' environmental literacy. Therefore, the hope for successfully promoting ESE in schools must go beyond completing basic accreditation programmes, but instead moving towards meaningful and integrated approaches that seek interconnectedness between stakeholders and content. While several whole-school approaches have allowed for strong interconnectedness between stakeholders and content, the following section discusses how inquiry-based and action learning align well towards promoting ecological change in both beliefs and actions.

Inquiry-based and action learning

Many scholars argue that inquiry-based learning (IBL), with a focus on action and/or problem-solving, is key to promoting sustainability (Aditomo, Goodyear, Bliuc & Ellis, 2013; Behrenbruch, 2012; Bosevka & Kriewaldt, 2020; Pretorius, Lombard & Khotoo, 2016; Warner & Elser, 2015). IBL places an emphasis on questioning and investigation (Aditomo et al., 2013), which are important for exploring ESE as it encourages teachers and learners to identify ecological problems and seek solutions through action-taking. This type of action learning requires higher-order thinking skills and allows learners to identify local problems and solutions (Pretorius et al., 2016). With a focus on place-based learning that relies on problem-solving and is action-based within the local community (see Chapter 6), IBL provides opportunities 'for critical exploration and interrogation of existing concepts of knowledge to know how they are created and utilised for both individual and community benefit' (Pretorius et al., 2016, p. 183). With an emphasis on local partnerships, Salter, Venville and Longnecker (2011) argue that for systematic change to occur in schools, schools need to adopt a whole-school, whole-system approach which incorporates local partnerships that allows for 'action competence and shared responsibility' (p. 150).

In addition to partnering with the local communities to encourage ecological action, Juntunen and Aksela (2013) argue that an IBL approach can evoke reflection on one's ecological attitudes and beliefs, making learning in subjects such as chemistry meaningful. In their study, they found that IBL encouraged positive attitudes towards chemistry because it supported 'individual decision-making processes and provokes socio-scientific discussion' (p. 160). They argue that IBL allows for education *through* science learning rather than education *in* science learning. In their study involving 105 upper-secondary students in three schools in Finland, Juntunen and Aksela (2013) claim their results revealed that IBL had a positive impact on their participants' attitudes towards chemistry and environmental literacy and their motivation to study. They argue that it was the focus on environmental and societal issues which related to their daily lives that allowed them to make meaningful connections to chemistry concepts. Therefore, IBL allows learners to explore content in a way that reflects the environmental issues that influence their daily lives.

By highlighting inquiry, action, problem-solving and meaningful change, an inquiry-based, whole-school approach provides opportunities for schools to interweave and embrace the diversity of stakeholders' beliefs and attitudes (including school leaders, parents, learners and community members) to create a culture that provides opportunities for reflection and action, both inside and outside the walls of the educational setting.

The teachers' stories: a three-dimensional analysis

Through the narrative inquiry of three-dimensional analysis, three distinct but interconnected themes were evident in the stories that Shaun and Alison told about their experiences of ESE in their schools. These three themes focused first, from a temporal dimension, on a shift in thinking over time about the importance of a non-linear approach to leadership; second, through a societal dimension, the need for a socially just ethos firmly established within the school; and finally, a place-based dimension which illustrated that a context that is conducive to 'big ideas' and in-depth inquiry enables the inclusion of sustainability across the curriculum.

A non-linear, blended approach to leadership

At the beginning of the first workshop, Shaun cautiously declared his feelings of 'luck' and good fortune in working in a school where ESE was a highly valued aspect of the school and its community. Shaun spoke in a quiet, almost embarrassed voice, suggesting:

> I'm probably the terribly lucky one here in that we can think big and we can implement integrated projects… and if we are not doing that, our Principal will come and let us know that we are not doing it right…
>
> *(Shaun, workshop one)*

At the start of the first two workshops, Shaun's voice was often heard reinforcing the need for strong leadership from the Principal, and how important it was for him, his colleagues and for his school community, such as when he said:

> If you have the Principal on board, you can do anything…
>
> *(Shaun, workshop one)*

When asked on another occasion if he had sought permission from the Principal to send an email providing ideas to *the entire staff* about Sorry Day and other Indigenous initiatives, he stopped and reflected on his actions and replied:

> I did [ask permission], but I wouldn't have had to. It was really just a, 'Hey Ester, as the Principal, I wanted to send this out, just because it might be sensitive, is it okay?' and she said, 'Yeah go for it' …
>
> *(Shaun, workshop one)*

In the following workshop, Shaun continued to explain how the whole-school approach worked in his school in terms of ESE, such as whole-school staff meetings aligning the *ever evolving* school philosophy with sustainability (Shaun, workshop two). However, what was interesting in Shaun's contributions to the iterative workshop conversations was that with each successive workshop, he became gradually more aware of how his own belief system and his strong environmental identity played a role in the school's successful enactment of ESE. In other words, his habitus shapes and is shaped by the institutional habitus of his school. Shaun's contentions about ESE are aligned with those of the social field. The Principal holds power within this social field and imparts 'cues' and 'logic' (Dalal, 2016, p. 235), which makes it possible for Shaun to align his practices to the rules of the game. This was evident, for example, when Shaun's stories shifted from merely drawing attention to the role of the Principal towards commenting on the combined importance of *both* a 'top down' and 'upstream approach' to ESE in the school, stating:

> If you're going to get schools on board there needs to be a top down approach… It needs to come down to the Principal saying, 'Right, this is what you need to do, go and do it.' That's right, and if we can combine that with the upstream approach… great!
>
> *(Shaun, workshop two)*

Again, later in the same workshop two, Shaun recounted strongly on the need to work with your own 'strengths' that you bring to your teaching, arguing that:

> I guess it's your beliefs that define you as an educator…Yeah, and the leadership has to be encouraging and I think that's what you're saying. So I agree, the leadership encourages you to go out there with your strengths and follow your strengths and different leaders can get stuck into making sure you iron out your weaknesses, which is important as well…
>
> *(Shaun, workshop two)*

Consistent with Shaun's stories and experiences, Barnes et al. (2018) and Henderson and Tilbury's (2004) earlier findings found that strong leadership and governance were key to a successful whole-school approach to ESE. Confirming Shaun's shift in thinking over time, these researchers also identified that support from the rest of the school community was an integral element to embedding ESE into the school and that the leader could not succeed alone.

In further re-evaluating his own role in the school, Shaun also started to contemplate the recruitment process at his school. Through this reflexive process, Shaun now realised the informal tour around the school he was taken on during his interview was purposefully arranged to ascertain his values, assumptions and belief systems. In retrospect, Shaun became increasingly aware that it was his explicit commitment to ESE, his ways of defining, describing and locating his environmental identity (Clayton, 2012) that saw him gain employment at this 'greenie' school rather than just a matter of 'luck' that he had initially suggested in the first workshop. Therefore, whether intentional or not, he was able to satisfy the rules of the game to gain legitimacy and voice, ultimately gaining cultural capital and power within this social field.

An ethical and socially just school culture

Both Alison and Shaun's stories often referred to the *culture* of their schools and how this informed their teaching, specifically around ESE. This reference to 'culture' is likened to institutional habitus that allows schools to position themselves in relation to stakeholders (Bartlett et al., 2015; Brickson, 2007), such as parents and the wider community. For example, in Alison's school, they held an annual Sustainability Expo in the school hall where the Grade 6 learners presented a visual model of some aspect of sustainability they were interested in, and then verbally explained the underlying sustainability principles to parents, teachers, other learners and community members, while at Shaun's school, sustainability was held on equal footing in the curriculum alongside literacy and numeracy by all teachers, in all classrooms across the whole school. Therefore, particular types of knowledge, in this case ESE, are negotiated and determined by the powerful players and social structures that organise a social field and these schools prioritised sustainability as an important type of knowledge. Both these early career teachers were intensively aware that their schools were known in each respective area (Shaun's inner-city school and Alison's outer semi-rural, both in high SES areas) as 'greenie' schools because of the inclusion of ESE into the curriculum. However, this term became contentious in a conversation between a parent and Alison, as depicted in her story at the beginning of this chapter, when she said:

> This is really awful; I had an excursion to CERES last year with the Grade 6s and I caught up with a student who hadn't yet paid and the parent responded, 'Oh yeah but what are you doing?' and I said, 'We're going to go CERES, that's our inquiry topic and they'll learn a few things about sustainability' and she said, 'Oh nah, that's okay, I'm just not really into that kind of [environmental] stuff'...
>
> *(Alison, workshop two)*

Alison was clearly disturbed by this sudden realisation that there were parents amongst their school community who did not share the same *environmentalism* philosophy as the teaching staff, leadership and majority of the local community. Alison continued her story, saying:

> I was like, 'Okay thank you, no worries.' Is that my role to sit on the parent and try and convince the parent that environmentalism is important, that looking after the earth is important...Yeah, well...Everyone has their issues and everything but I was really shaken up by that...That was the first time I've ever come across that in this school.
>
> *(Alison, workshop two)*

In thinking more deeply about this ethical dilemma during workshop two, Alison questioned her response and re-considered what she could have done to rectify this situation, suggesting instead:

> Maybe it was misinformation with one parent so maybe you can open up that conversation to get them into the classroom and say, 'This is what we're discussing, this is what the students have been working on.' So get that conversation happening with the students saying why - if they come in and see their work and the students have to re-teach their parents how

to do a fraction or whatever it is, I suppose you can take that approach as well with environmentalism.

(Alison, workshop two)

Conversations around *environmentalism* seen as merely a *greenie fad* continued amongst the participants following on from Alison's story, as they pondered the challenges of behavioural change within educational settings and local communities. Similarly, Evans, Whitehouse and Gooch (2012) found in their Far North Queensland survey of schools and teachers that being positioned as a 'greenie' was seen as 'socially unacceptable' and therefore a barrier to adopting ESE into their schools (p. 124). In a revealing statement, some of the participants in the current study disclosed that although their schools had 'signed up' with Centre for Education and Research in Environmental Strategies (CERES) and Dolphin Research Institute (DRI), they were not always fully committed to sustainability within their school culture. Therefore, as Goldman et al. (2018) had found in their study on the different levels of accredited Sustainability programmes, this meant that the deeper learning about sustainability did not always filter down to real understandings amongst the learners.

However, it was Shaun's stories emphasising the *ethical perspective* rather than focusing on the *facts and statistics* of sustainability that lifted the conversation back to more hopeful endeavours. Shaun felt discussions, experiences and learning around social justice issues in sustainability were 'more deeply personal' and meaningful for him and significantly, for the learners in his classroom. This was more so as these opportunities were fine-tuning his identity formations in ways that allowed him to ground himself and give meaning to his beliefs. Following this statement, Shaun appealed for more emotional engagement with learners by thinking about the sort of ethical problem-solving they needed to consider now and into the future around sustainability, and said:

> ...in terms of being a leader here you want them to take some ownership. So we've started the talk here, and they've noticed it [e.g., litter in the playground] and they've noticed it's a bit of a problem, now it's your job not to do it for them but help them learn what they can do about it.

(Shaun, workshop three)

Shaun's suggestion of enabling his learners to become more aware of their ethical responsibility towards their environment by noticing changes, potential social justice issues and then working on problem-solving skills was agreed upon by all participants. This was evident, for example, in another story Shaun told of how one of his learners had noticed a change over time in Shaun's sustainable practices, saying:

> I'm thinking of one conversation I had with a student a couple of weeks ago. So at the start of the year I was walking into the classroom and I had my whole pile of stuff that I needed, and now I've just got my computer

and he's like, 'Shaun, where's all your stuff? You're not carrying as much paper as you used to' and I'm like, 'But you should see all the tabs that I've got open on my computer. I'm just not printing them. I've got them all open online'... and that's one little thing that he noticed. So just challenging the kids, they're like, 'We need to print something' and you can say, 'But you've probably got that online, so you don't need to print it'...

(Shaun, workshop three)

After listening to this story, the collective group also agreed (as Barnes et al. (2018) have contended) that this learning experience was not something that could be quantitatively measured by NAPLAN assessments, and so tended to be missing or ignored in the standardised curriculum of many schools. In contrast, ESE in these schools was realised as powerful knowledge within their social field which positioned these teachers as powerful and knowledgeable actors.

An environment conducive to 'big' questions and new ideas

Contrary to Dyment and Hill's (2015) findings that pre-service teachers have a narrow view of ESE in the classroom, Shaun's personal and emerging professional eco-identity was well entrenched in the broad spectrum of sustainability while at university. It was an integral part of his habitus, influencing his ways of thinking and doing—which just happened to align well with his school's institutional habitus. As a pre-service teacher, Shaun was already thinking about how closely aligned ESE and IBL were when he was working on a project with local school children during his sustainability unit. Shaun told the community of practice group how it was a defining moment for him when he realised the importance of providing opportunities for learners to think about and ask 'big' questions relating to sustainability, saying:

> I think it was when we did sustainability and working with some of the kids in the 'local' school. These big questions would be asked by the kids and we'd start exploring topics and it was the answers that we didn't have. Like, there were all these identified problems we didn't have solutions for in terms of sustainability and the environment and it was kind of like those moments where I went okay, I don't have the answers, they aren't obvious, this is clear to me that we are engaging with critical thinking. This is where we are giving the kids the opportunity to think about the future... it really felt like we were stepping forwards...
>
> *(Shaun, Workshop one)*

Shaun's burgeoning understanding that teaching is not just about the transmission of knowledge, but rather a shared experience with learners through authentic inquiry, questioning and problem-solving, was an important development in his professional identity. Given his strong personal and professional eco-identity, it is not surprising then that Shaun applied (and was subsequently employed) to teach at a school that was perceived within Victorian educational networks and the local community as a 'greenie' school. As early career teachers must position themselves within institutional practices and norms, finding a strong alignment between one's professional and moral beliefs (Colegrove & Zuniga, 2018) with that of their school allows for a strong sense of agency.

At the beginning of the first workshop, Shaun had declared to the group that he was the 'lucky one' to be teaching in a school that valued ESE as a matter of course; so, unlike many of the other early career teachers in the group, he did not need to push back against the school culture to systematically include sustainability in the curriculum. In making this statement, Shaun was acknowledging the influence of an educational context that encouraged and enabled his personal and professional eco-identity to unite effectively in his teaching. In contrast, Evans et al. (2012) found that cultural, social and political elements of the educational context can present a barrier through the 'resistance by school members to adopt a whole school approach to change,' and so prevent the enactment of ESE in the classroom (p. 124).

A critical element of Shaun's story was his school's pedagogical decision to use learners' ideas through IBL as the basis of all their curriculum planning, teaching and learning. The continuation of his story links IBL with ESE in a whole-school approach, saying:

> So, we start with an inquiry and we move through different questions… And now there are three groups from that one inquiry. One group is looking at Indigenous perspectives, there's a group looking at sustainability and a group looking at historical context. And we've begun to look at how these are all wrapping around each other and linking as well, but everything starts with that inquiry. We, all the teachers, sit at planning meetings and we talk about what the kids are interested in and then we sort of ask ourselves what are the next steps through that…
>
> *(Shaun, workshop one)*

As with Juntunen and Aksela's (2013) study, Shaun's stories illustrated that the intentional use of IBL can evoke reflection of one's own ecological attitudes and beliefs. As an early career teacher, Shaun recollected that he had left university concerned that he would need to strictly abide by a NAPLAN-directed curriculum with a Principal and parent body who expected education to focus purely on numeracy and literacy. However, very quickly after his initial introduction to his school, Shaun found it 'amazing' that there was 'so much more curriculum' through the inclusion of ESE that could be 'ticked off in a more meaningful context' through an integrated, inquiry-based pedagogical approach (Shaun, workshop three). In the community of practice workshops, Shaun consistently encouraged other early career teachers to try to teach in this integrated way, hoping their principals and fellow teachers would in time see the clear benefits to the learners, to the school and to the community.

> ## Box 7.2 Extract from Alison in her reflective response
>
> I have really enjoyed our sessions. It has been fantastic to be able to share ideas, resources and network with everyone involved, and learn more about the environmental organisations within Melbourne and its surrounds. Being able to share common challenges and limitations with peers, highlighted the fact that we are bound by common restraints and facing similar challenges. It made me realise that it takes quite a bit to create change, at a small and larger level. I am extremely grateful that the school I am employed at has an interest and commitment to environmental and progressive educational opportunities.
>
> *(Alison, Reflective response)*

Alison also referred to the significance of teaching in a school that valued and promoted ESE, a school environment that was committed and conducive to enabling 'environmental and progressive educational opportunities' for the learners and the teachers (Alison, reflective response). At the end of the project, we had invited the early career teachers in our community of practice group to make one final contribution to the project. The invitation was to provide an email response of their reflections on how the project and the group may have informed their current and future teaching. Subsequently, Alison's reflective response (see Box 7.2) represented what we as researchers had hoped was a positive outcome for the project, and illustrated the impact of the educational context on this early career teacher's capacity to enact ESE in the classroom and within her school.

Alison's comments highlight the need for early career teachers to have some alignment between their professional and moral contentions, the confluence of their identities and that of their school—as this allows for an important sense of legitimacy and power that provides teachers with agency. Alison and Shaun felt they had agency because they understood the institutional norms and practices of their school and these practices aligned with their own.

Looking to the future—hopeful applications

> You're forever reflecting and adapting to whatever students you do have, but I think to be able to see the potential in the students and to be able to empower them, I think that's really something to keep in mind whatever way you are incorporating environmentalism into your subjects. I think that's something that throughout teaching – so what do they walk away with? Do they walk away with questions or... just being able to maybe ignite a little bit of empowerment in a few then that's something that gets you back to school day after day, I suppose.
>
> *(Alison, workshop two)*

This chapter has provided empirical evidence through the stories of two early career teachers who taught in two *greenie* primary schools where ESE was an integral element of their whole-school approach to their curriculum. It has illustrated through the teachers' stories that a blended, non-linear 'up and down' approach to leadership is needed, not just top-down from the Principal, though the Principal's support is critically important as a starting point. Their stories have also indicated that the inherent structure of the school culture impacts on the early career teachers' prerogative to make curriculum suggestions and pedagogical decisions; be part of a team that thinks collectively about what the learners are interested in learning; and whether or not socially just elements of ESE (such as thinking about intergenerational equity) underpin the IBL experiences. The analysis of these stories has also shown the significance of context (geographically, culturally, politically and/or socially) and how particular 'places' appear to be more conducive to enabling ESE than others. Alison's final story extract above portrays how important her ESE work is for her developing sense of self as an effective teacher in assisting the growth of her learners.

References

1 (One) million women. (2015, April 16). Labels are not for people! Remove the stigmatised terms "greenie" and "treehugger." Retrieved from https://www.1millionwomen. au/blog/labels-are-not-people-remove-stigmatised-terms-greenie-and-treehugger/

Aditomo, A., Goodyear, P., Bliuc, A., & Ellis, R. A. (2013). Inquiry-based learning in higher education: Principal forms, educational objectives and disciplinary variations. *Studies in Higher Education, 38*(9), 1239–1258.

Australian Sustainable Schools Initiative & Australia. Department of the Environment, Water, Heritage and the Arts. (2010). *Australian Sustainable Schools Initiative (AUSSI).* Canberra: Department of the Environment, Water, Heritage and the Arts.

Barnes, M., Moore, D., & Almeida, S. (2018). Sustainability in Australian schools: A cross-curriculum priority? *Prospects, 48*(1), 1–16. doi:10.1007/s11125-018-9437-x

Bartlett, J., McDonald, P., & Pini, B. (2015). Identity orientation and stakeholder engagement: The corporatisation of elite schools. *Journal of Public Affairs, 15*(2), 201–209.

Behrenbruch, M. (2012). *Dancing in the light: Essential elements for an inquiry classroom.* Rotterdam, The Netherlands: Sense Publications.

Bosevska, J., & Kriewaldt, J. (2020) Fostering a whole-school approach to sustainability: Learning from one school's journey towards sustainable education. *International Research in Geographical and Environmental Education, 29*(1), 55–73. doi:10.1080/10382046. 2019.1661127

Brickson, S. (2007). Organizational identity orientation: The genesis of the role of the firm and distinct forms of social value. *Academy of Management Review, 32*(3), 864–888.

Clayton, S. (2012). Environmental identity: A conceptual and an operational definition. In S. Clayton (Ed.), *Oxford handbook of environmental and conservation psychology* (p. 164–180). Oxford: Oxford University Press.

Colegrove, K., & Zuniga, C. (2018). Finding and enacting agency: An elementary ESL teacher's perception of teaching and learning in the era of standardised testing. *International Multilingual Research Journal, 12*(3), 188–202.

Dalal, J. (2016). Pierre Bourdieu: The sociologist of education. *Contemporary Education Dialogue, 13*(2), 231–250.

Dyment, J. E., & Hill, A. (2015). You mean I have to teach sustainability too? Initial teacher education students' perspectives on the sustainability cross-curriculum priority. *Australian Journal of Teacher Education, 40*(3). doi:10.14221/ajte.2014v40n3.2

Evans, S., Whitehouse, H., & Gooch, M. (2012). Barriers, successes and enabling practices of education for sustainability in Far North Queensland schools: A case study. *The Journal of Environmental Education, 43*, 121–138. doi:10.1080/00958964.2011. 621995.

Goldman, D., Ayalon, O., Baum, D., & Weiss, B. (2018). Influence of 'green school certification' on students' environmental literacy and adoption of sustainable practice by schools. *Journal of Cleaner Production, 183*, 1300–1313.

Griffin, D., & Chan, T. (2015). PPBES: How one school district goes green. In T. Chan, E. Mense, K. Lane, & M. Richardson (Eds.), *Marketing the green school: Form, function and the future* (pp. 258–266). Hershey, PA: IGI Global.

Henderson, K., & Tilbury, D. (2004). *Whole-school approaches to sustainability: An international review of whole-school sustainability programs.* Report prepared by the Australian Research Institute in Education for Sustainability (ARIES) for The Department of the Environment and Heritage. Australian Government. Retrieved from http://aries. mq.edu.au/projects/whole_school/files/international_review.pdf

Juntunen, M., & Aksela, M. (2013). Life-cycle thinking in inquiry-based sustainability education: Effects on students' attitudes towards chemistry and environmental literacy. *Centre for Educational Policy Studies [CEPS] Journal, 3*(2), 157–180.

Koç, A., & Bastas, M. (2019). The evaluation of the project school model in terms of organization sustainability and its effect on teachers' organizational commitment. *Sustainability, 11*, 1–23.

Merriam-Webster Dictionary. Retrieved on 2 June, 2020 from https://www.merriam-webster.com/dictionary/tree%20hugger

Metzger, A. B. (2015). Green school frameworks. In T. Chan, E. Mense, K. Lane, & M. Richardson (Eds.), *Marketing the green school: Form, function and the future* (pp. 1–14). Hershey, PA: IGI Global.

Pretorius, R., Lombard, A., Khotoo, A. (2016). Adding value to education for sustainability in Africa with inquiry-based approaches in open and distance learning. *International Journal of Sustainability in Higher Education, 17*(2), 167–187.

Salter, Z., Venville, G., & Longnecker, N. (2011). An Australian Story: School Sustainability Education in the Lucky Country. *Australian Journal of Environmental Education, 27*(1), 149–159.

Sustainability and Environmental Education [SEEd]. (2020). Sustainable schools. Retrieved from: https://se-ed.co.uk/edu/sustainable-schools/

Uren, H., Dzidic, P., Roberts, L., Leviston, Z., & Bishop, B. (2019). Green-tinted glasses: How do pro-environmental citizens conceptualize environmental sustainability? *Environmental Communication, 13*(3), 395–411. doi:10.1080/17524032.2017.1397042

Warner, B., & Elser, M. (2015). How do sustainable schools integrate sustainability education? An assessment of certified sustainable K–12 schools in the United States. *Journal of Environmental Education, 46*(1), 1–22. doi:10.1080/00958964.2014.953020

8

ESE IN EARLY CHILDHOOD EDUCATION

'We do lots of little things … now'

Box 8.1 Extract from Jessica in workshop two

I work at [beach side] Early Learning Centre and I'm the educational leader. I think lots of people look at me for what to do, especially with programming and what they're doing in their rooms. That's fine with me but I'm more of everyone has a say and we do it all together, not for me to do it all. So with environmental [education] I'm interested in it but we don't really do it at the moment…

(Jessica, Early Childhood Teacher, outer eastern beachside area,
workshop two)

Introduction

Of the 11 teachers who responded to our invitation to join the project, two were early career kindergarten teachers: Jessica, a kindergarten teacher with the position as the educational leader for a beachside long day-care centre; and Bonnie, a kindergarten teacher in an inner-city long day-care centre who was the sustainability educator for the local government organisation where she worked. Unfortunately, due to work commitments, Bonnie was only able to attend one of the workshops, while Jessica attended most of the community practice workshops. While literature has shown that a purposive focus on sustainability has been 'slow' to be incorporated into early childhood education (Davis, 2009; Elliott & Davis, 2018), the two kindergarten teachers in this study had vastly different lived experiences of Environmental and Sustainability Education (ESE) from both ends of the continuum in their early childhood workplaces.

The provocation at the beginning of this chapter signposts one of the key arguments highlighted in this chapter regarding the tensions around leadership in sustainability, which Jessica suggested created challenges in her workplace. This chapter provides a three-dimensional narrative inquiry analysis (Clandinin & Connelly, 2000) of the stories the early career kindergarten teachers told over the course of the study. They included the temporal dimensions (the historical traditions in early childhood education); societal dimensions (educational policy shifts, notions of educational leadership; and the image of the young child as a change agent) and the place-based dimensions (in early childhood settings and in relation to place-based learning) that informed and influenced Jessica's and Bonnie's experiences and the stories they chose to tell at that time and place.

This analysis affords an insight into how both teachers positioned themselves within the field of early childhood education, particularly as they negotiated what it looked like to embed 'sustainable practices' into their early childhood settings and practices. In an additional layer of analysis, the chapter also shows how these conflicting dimensions had simultaneously impacted on one teacher's emerging professional and personal identity (or habitus) in relation to EC ESE within her early childhood centre (the field). As explained in the methodology in Chapter 5, we had started our study with an (unrealistic) expectation that all participants would be motivated through our workshops to design and activate a sustainability project in their own educational setting. However, what we had not envisaged was how powerful the contextual constraints could be in hampering the progress of EC ESE in an early childhood setting.

The first section in this chapter discusses the 'slow' introduction of sustainability into early childhood education, and how its emergence has taken time to become accepted within the field due to the ongoing influence of historical informants. Three key barriers to the establishment of ESE in early childhood settings are explained; so too, the capacity of young children to act as sustainability 'change agents.' Following this overview of the literature, the chapter provides storied evidence of the contextual dimensions surrounding the early career kindergarten teachers in the study and how these teachers negotiated the field in light of their habitus and their confluence of identities. To conclude the chapter, hopeful applications are offered to further the implementation of EC ESE into early childhood education.

Sustainability in early childhood education

In Chapter 1, we discussed the multiple iterations of the terminology used to describe the inclusion of sustainability into education. Our decision as researchers has been to use the blended term Environmental and Sustainability Education (ESE), to take into account all the underlying ideologies, principles and practices involved in both areas of environmental education and of education for sustainability (EfS). As many significant researchers have done before us, this chapter takes this term further and includes Early Childhood (EC) with Education for Sustainability ((Davis, 2009, 2015; Elliott, 2003; Elliott & Davis, 2018; Huggins & Evans, 2017) and/or Environmental and Sustainability Education (ESE) (Wals & Benavot, 2017), and will be represented as EC ESE throughout this chapter.

While Davis (2015) has lamented that sustainability is still a 'confused and contentious topic' (p. 9), she concludes that sustainability is 'essentially an issue of social justice and fairness' rather than just learning 'about' and 'in' the environment (p. 10). For Davis (2015), sustainability is about 'how we live our lives now and into the future' (p. 3).

Further to this notion of equity and social justice in relation to human impact on the environment, Elliott and Davis (2018) and more recently, Elliott, Ärlemalm-Hagsér and Davis (2020) urge us to consider the next generation who will be severely impacted by climate change. Similarly, Cutter-Mackenzie and Rousell (2019) argue that there is a critical issue of 'generational injustice,' whereby 'children and young people are currently inheriting social and ecological problems which they had very little part in creating' (p. 90). However, Davis (2015) has long argued that young children should not be seen as 'victims' nor positioned as 'saviours of the world' who need to be educated in order to take care of the planet, but as 'creative problem solvers' who have opinions and ideas to offer about sustainability (p. 16). This acknowledgement of children's agency aligns well with both the UN Declaration on the Rights of the Child (1989) Article 12, whereby children have the mandated right to voice their opinions on matters that pertain to them and the sociology of childhood paradigm that puts children forward as capable 'social actors' in their own lives (Dahlberg & Moss, 2005).

Drawing on Davis' (2015) definition of Early Childhood EfS, our own understanding of EC ESE aligns well with the powerful language of advocacy included here, where she states:

> Early childhood education for sustainability can be described as the enactment of transformative, empowering and participatory education around sustainable issues, topics and experiences within early childhood contexts.
>
> *(p. 22)*

The purposeful use of the words 'transformative' and 'empowering' are fundamental to EC ESE with an inherent pedagogical approach that 'values, encourages and supports children to be problem solvers and action takers in their own environments' (Davis, 2015, p. 23) through intentional teaching (Cutter-Mackenzie, Edwards, Moore & Boyd, 2014) towards sustainability dispositions, such as social justice, fairness and empathy (O'Gorman, 2020).

Individually and collectively, over the past two decades at least, significant research and publications by Davis (cf: 2009, 2015) and Elliott (cf: 2003, 2016) have illustrated the importance of including sustainability in early childhood education. However, despite this extensive work, both authors have found significant resistance within the early childhood profession towards teaching about sustainability in early childhood education (Davis, 2009; Elliott, 2003). This was especially evident in Elliott's (2003) *Patches of Green* report, which highlighted the enormity of the push back from early childhood educators to incorporate sustainability into their educational programmes with young children, whilst Davis' (2009) early work focused attention on the resistance to sustainability in research and pedagogical practices in ECE.

Despite the substantial work these and other researchers have done in the field of EC ESE, (cf: Hirst, 2019; Mackay, 2012; Siraj-Blatchford & Huggins, 2015), in reality, early childhood teachers have been historically 'slow' to intentionally include sustainability into their curriculum (Davis, 2015; Elliott & Davis, 2018). Unfortunately, Elliott and Davis (2018) are still concerned that the 'uptake of EC EfS both by educators and researchers has been tardy, an untenable situation given now pressing global concerns' (p. 165). However, Elliott and Davis (2018) report a slight 'positive surge in uptake since EC EfS' was recognised in the UN Decade of Education for Sustainable Development at the end of 2014 (p. 174).

Similarly, Huggins and Evans (2017) identified a small shift in thinking and implementation of EC ESE, with some early childhood teachers starting to move away from only 'teaching children facts about the environment to educating children to act for change' (p. 3). However, Huggins and Evans also identified significant barriers which prevented early childhood teachers from wanting to engage with ESE. According to their findings, Huggins and Evans (2017) found these barriers were based on 'unquestioned assumptions' about what 'should' be incorporated into the 'core functions of early childhood education' (p. 4). Three key barriers were identified against the uptake and implementation of ESE into early childhood education:

1. A lack of certainty in 'how to move beyond environmental education into EC EfS';
2. Early childhood teachers believe children should be protected and sheltered from the 'harsh realities' of life for as long as possible, seen to be the 'key function of early childhood education';
3. Early childhood teachers do not want to enter into a debate about sustainability, considering it to be an 'overly politicised' activist statement rather than a pedagogical decision to be made (Huggins & Evans, 2017, p. 4).

The following section explains these three identified barriers in more detail, with past historical informants combined with contemporary attitudes as potential causes for early childhood educators to block the inclusion of EC ESE into their educational programmes.

Barriers preventing the inclusion of sustainability into ECE

Long-held beliefs and values that have formed the cornerstone of early childhood education can be seen to have originated from early child development theorists and philosophers from the 18th and 19th centuries (Platz & Arellano, 2011). Historical theories based on the 'innocence of children' attributed to Rousseau in the 18th century (James, Jenks & Pence, 1998, p. 13), and the 19th century image of the 'young child in nature' who needed to be protected from the reality of the adult world by being ensconced within Froebel's walled garden (Dahlberg, Moss & Pence, 2007) are recognisable and dominant tenets that still hold strong sway in the pedagogical practices of contemporary early childhood education (Dahlberg & Moss, 2005). As a consequence, Elliott et al. (2020) argue these 'theoretical legacies and practice histories dating back to the 19th century, along-side more recent calls for young children to learn to love nature' have given rise to an increasing 'tendency to supplant nature play as education for sustainability' in early childhood education (p. xxv).

While researchers and writers on EC ESE still advocate for the importance of outdoor natural play spaces for the health, well-being and development of young children (cf: Elliott & McCrea, 2016; Kahn, Weiss & Harrington, 2019; Moore, Morrissey & Robertson, 2019), they also argue that outdoor play alone is 'insufficient' for encouraging dispositions, conceptual understandings and skill development needed for EC ESE (Elliott, 2003; Elliott & McCrea, 2016; Luff, 2018). Elliott (2003) claims that a focus purely on outdoor play is 'only one part of what makes up the full scope' of early childhood sustainability (p. 2); whilst Elliott and Young (2016) have further problematised early childhood teachers' tendency towards a 'nature by default' approach to pedagogical decision-making. They argue this approach could 'potentially thwart [...] a fuller transformative engagement with sustainability' and may not enable children's inquiry into the deeper aspects of sustainability' (Elliott & Young, 2016, p. 58). However, in reference to Huggins and Evans' (2017) finding about the perceived politicisation of sustainability, researchers have found those seen to overtly support sustainability '**for** the environment' rather than simply caring '**about** the environment' (Davis, 2015) are frequently labelled as 'activists' (Uren, Dzidic, Roberts, Leviston & Bishop, 2019). Uren et al. (2019) claim that while many Australians 'aspired to be green' and claimed to 'hold a sustainability identity' (p. 395) they did not want to be seen as 'activists' in relation to sustainability, and subsequently 'distanced themselves from social activism' (p. 395).

Given the key barriers to early childhood teachers' engagement with ESE that Huggins and Evans (2017) have identified, a clear mismatch is evident between deeply entrenched early childhood philosophies and pedagogies based on concepts of childhood innocence, protection and nature in contrast with the idea of teaching young children to be active 'change agents' for sustainability.

Young children as 'change agents' for ESE

It is timely to highlight the unique position of young children in the ESE space. This is particularly so in relation to starting sustainable ways of living as early as possible so that children can 'create' sustainable practices from a young age rather than having to 'change' their behaviour later in life (Elliott, 2003; Lemmon, 2020; Siraj-Blatchford & Huggins, 2015). It also highlights the need to acknowledge the power and agency in young children, seeing them as capable of leading transformational change. In Greta Thunberg's book (Thunberg, 2019), she argues that every child has the capacity, capability and agency to make a difference—it is just a question of adults ensuring that they have the opportunities to harness this potential. In a personal communication with Julie Lemmon (29 February 2020) from the Clarendon Children's Centre, South Melbourne, during the Stephanie Alexander Kitchen Garden (SAKG) Early Years programme launch, Julie reinforced young children's agentive capacity in sustainability. She emphasised the importance of young children being given the opportunity to 'create' their sustainable living behaviour by saying, 'we are cooking for the plate not for the bin' (SAKG Early Years launch, February 2020).

Similarly, the UNESCO report on Climate Change Education for Sustainable Development (2010) also advocated for starting sustainable practices early, stating that 'instilling climate change awareness and understanding at a young age is ultimately the best way to change behaviours and attitudes' (p. 5). In a valuable addition to this argument, Davis (2010) has provided an explanation on the significance of EC EfS as a 'catalyst for change' through the 'power of the very young,' stating:

> Not only can young children learn about environmental issues; they can, and do, take action to change their behaviours in both the educational setting and at home. Furthermore, the adults around them – their teachers and parents - also learn and act for the environment as a result of the curriculum decisions that are made… [it] is about the power of early childhood education for sustainability as a catalyst for change, and the power of the very young as agents of change for sustainability.
>
> *(p. 22)*

Core to this argument on the 'power' of the 'very young as agents of change' is the phenomenon Chaudhary (2018) has identified as 'reverse socialization,' where 'parents are socialised by their young children' into behaving in particular ways (p. 347). Stuhmcke (2012) also found that young children were able to act as sustainability 'change agents' by influencing their parents' shopping habits. For example, this was seen when young children were searching for recycling symbols and reduced packaging while shopping with parents, telling their parents that it is 'better for the environment' (p. 136). Anecdotally, young children have been reported to be the 'change agents' responsible for the success of the Victorian, then Australian and eventually the global seat belt campaign during the 1960s when they 'pestered' parents into remembering to do up their seat belts.

The notion of young children's capacity as 'change agents' is frequently highlighted in contemporary early childhood literature, with researchers eager to dispel the myth that young children are not capable of discussing 'harsh' topics, such as the impact of humans on the earth (Davis, 2015; Duhn, 2012; Edwards, Moore & Cutter-Mackenzie, 2012; Elliott & McCrea, 2016). For example, Duhn (2012) acknowledged that traditionally early childhood teachers have 'shielded' children by staying away from complex topics such as sustainability; however, she argued that teachers now need to work with children, families and the community so that young children have the opportunity to learn about EC ESE in meaningful ways (p. 20). Similarly, Elliott and Davis (2018) have continued to strongly advocate for 'as early as possible' to start discussions around sustainability, which contrasts with the 'as late as possible' argument by 'many' early childhood teachers reported in Huggins and Evans' (2017) study.

Advocating for an 'as early as possible' attitude was exemplified on an international platform ten years ago at the Organisation Mondiale pour L'Education Pre-Scolaire (OMEP) World Congress on EC EfS in Sweden in 2010, where the whole focus of the Congress was on young children's perspectives on sustainability. The notion of young children's role in 'caring' or 'cleaning up' the world was foregrounded in the proceedings, including the words of a five-year-old Irish boy, who was reported as saying, 'We're saving the world for later.' Davis (2015) claims there is a 'blind spot' about what children already know, experience and understand about their world, stating 'No matter how much we may wish it to be otherwise, children are exposed to, impacted upon and affected by real-world issues' (p. 23). Davis' (2015) comments have become especially relevant over the Australian summer of 2020 with the devastation of the bushfires, the thick smoke shrouding large areas of the city and country alike, and the scenes of injured and/or dead wildlife seen daily through the media. Charlotte, the four-year-old granddaughter of one of the authors, was so impacted by the thick smoke hovering over her inner-city Melbourne home and kindergarten for many days that the bushfires and the dying animals became a topic of constant conversation. In response to this, Charlotte and her family organised a donation to 'Save the Koalas' as a way for Charlotte to feel she was 'helping the koalas in the bush fires' she said. Burke (2020) asserts that 'policymakers that demand educators and adults "stop worrying the children" with their language and analysis of climate change have got it wrong' (p. 17). Instead, Burke (2020) argues that young children are suffering from 'eco-anxiety' because of what is seen to be happening in the environment in reality (such as the smoke and poor air quality during and after the 2020 Australian bushfires), and that therefore this anxiety needs to be talked about and validated, not suppressed through false platitudes (p. 17).

However, it should be noted that as Mackay (2012), Davis (2015) and Elliott and Davis (2018) have consistently argued, young children should not be viewed as the 'saviours' of the world and that this increasingly heavy burden should not be placed on their shoulders. Rather, EC ESE is equipping young children with the critical literacy skills, dispositions and awareness of what can be done to work towards sustainable living, with Mackay (2012) stating:

> Teachers, who understand the importance of democratic processes when working on environmental problems and solutions, take the position that the child already has agency and has a contribution to make towards creating a better future.
>
> *(p. 474)*

Despite the increasing acknowledgement of children's capacity to discuss what is happening in and around their lives (Dahlberg & Moss, 2005), there is often a palpable tension within early childhood education that the topic of 'sustainability' is still not something that young children should be concerned about (Boyd, 2019; Huntley, 2020). Similar to Huggins and Evans' (2017) findings, Boyd (2019) suggests that many teachers and parents see an early childhood setting as a 'safe place' where children are sheltered from the 'more unpleasant aspects of life' (p. 984). One way that Boyd (2019) has found to successfully circumnavigate parental and teacher resistance to ESE has been through place-based learning where young children's 'ecological identity' can be nurtured and developed (p. 993). Boyd provides a UK-based example of place-based learning with young children who were given the opportunity to explore their local beach as part of an increasing focus on ESE. Boyd's study illustrates how starting 'small' with children's rubbish collection on their 'local beach' can expand the children's thinking and problem-solving into other sustainable practices, such as saving paper and recycling strategies. The multiple and multifaceted barriers (such as teachers wanting to protect children from the realities of life; and the perception that sustainability is overly political to discuss with children) and potential affordances (such as children's capacity to be change agents and place-based learning beyond being 'in' nature) have played an influential role in the way sustainability has been perceived, positioned and/or enacted in early childhood educational settings. New ways of looking at the underlying reasons for the enactment (or lack) of EC ESE will be examined through the analysis of the early career teachers' lived experiences in the following section.

The teachers' stories: a three-dimensional analysis

The following sections move into the three-dimensional analysis for this chapter, starting with the temporal dimension around historical informants still influencing early childhood teachers' attitudes to sustainability, and how there is a slow uptake of ESE in the field. Next, societal aspects connected with ESE and early childhood policies and educational frameworks are examined, and the issue of young children as agentive 'change agents' is raised. In the final dimension, place is highlighted as a significant dimension from a workplace stance as well as the context for local place-based learning for ESE initiatives.

The slow uptake of ESE: *'We don't really do it at the moment'*

Although the analysis of the temporal dimension of the kindergarten teachers' stories identified evidence of past historical informants, it was more often implied in the underlying meanings of the stories that were told. For example, this was evident in a story that Jessica told in workshop two, where she referred to an interest in but minimal engagement with *environmental education science*, when she said:

> So... with environmental education science I'm interested in it but we don't really do it at the moment and I think I need to become more interested in it and more passionate about it and then hopefully everyone else will follow as well...
>
> *(Jessica, workshop two)*

In this story extract, Jessica's use of 'we' rather than 'I' suggests that Jessica aligns herself with a collective view that ESE is not seen as a priority in early childhood education. In other words, Jessica's ways of thinking and doing, or her habitus, are shaped by the field. This highlights how Jessica, in contrast with other participants in the study, does not see ESE as an integral part of her everyday doing and thinking (habitus), but that rather her understanding of *environmental education science* heavily relies on the historical informants that have shaped the field of early childhood education over time. This is interesting because Jessica had been identified as a pre-service teacher who had demonstrated a strong interest in sustainability at university. However, from Jessica's stories it appeared that the documented resistance to include ESE in early childhood education (Davis, 2009; Elliott & Davis, 2018; Mackay, 2012) was so pervasive amongst her fellow educators in her early childhood centre that it had filtered into Jessica's pedagogical planning and teaching as well. Jessica's comment 'hopefully they will follow as well' suggests she was conscious of her colleagues' push back against implementing ESE. As such, it is important to understand and appreciate the overarching milieu of the context at this time and in this place, and how this had impacted on Jessica's limited capacity to introduce ESE programmes that were not seen as important by her colleagues and the centre management.

Further to the literature that refers to the 'slow uptake' of ESE in early childhood education (cf: Elliott & Davis, 2018), Jessica's comment 'we don't really do it at the moment' confirms this phenomenon. Her story implies not only hesitancy, but also a sense of uncertainty around how *it* (neither sustainability nor *environmental education science*) could be integrated into the centre's curriculum. Drawing on Huggins and Evans' (2017) findings, a lack of certainty in 'how to move beyond environmental education into EC EfS' can also be seen in Jessica's stories when she said she needed to be 'more interested in it and more passionate about it' before it was possible to incorporate any aspect of sustainability (or *environmental education science*) into the curriculum at her centre. In this way,

Jessica was setting herself abstract and unachievable goals in her role as Educational Leader before having to address any other educators' lack of sustainable practices in their programmes. Jessica was also distancing herself from her strong environmental identities that were evident during her studies at university. This brings to light the dissonance and uncertainties teachers face when their existing identities do not match up with their perceived professional roles. Under such circumstances, teachers might 'play down' their identities in order to better assimilate with the dominant culture of the workplace.

However, over the course of the project, Jessica's engagement with some elements of ESE was starting to emerge in her stories. This was evident, for example, in workshop five, when Jessica spoke of taking the children in her room 'outside every day' for their lunch and teaching the children to be responsible for their own rubbish collection, as seen here:

> We started this term and we eat outside every day, and we have a Magpie that comes so I use that as an 'in' with the kids and talk about how if we leave rubbish it'll go and eat the rubbish and it'll get sick and we can't help it get better because it's wild…
>
> *(Jessica, workshop five)*

Jessica's reference to being 'outside' and her 'use' of a Magpie as a way to motivate the children's collection of rubbish could be seen as engendering empathy and care of more-than-human species (O'Gorman, 2020). Alternatively, it could also be seen as a conflation of the terms 'nature' and 'sustainability' as synonymous concepts which researchers have found is a common misconception in early childhood education (cf: Elliott et al., 2020). The powerful influence of dominant discourses on the way early career kindergarten teachers perceive their role and their work in early childhood education is illustrated when you reframe Jessica's stories through the lens of the strongly held historical informants of innocence, protection and nature (Duhn, 2012; Mackay, 2012). Further perceptions around early childhood policy documents and the shifts that have occurred in relation to ESE will be discussed in the next section and how this can be seen through Jessica's lived experiences.

Policies shifts in early childhood education and ESE

In the societal dimension of analysis, a number of key themes, patterns and contrasting silences were identified within the stories the early career kindergarten teachers told. These findings focused on a range of policy documents relating to sustainability in early childhood education, while young children's capacity to be 'change agents' was brought to the foreground in many conversations. What is interesting in the societal analysis of Jessica's stories was her perception of the differences between what had impacted on her engagement with ESE compared with the early career primary school teachers. During workshop two, Jessica revealed that she felt an absence of visible engagement with ESE in early childhood education (and indeed, in her centre) was not based on a lack of direction from early childhood curriculum policies. Nor did she feel it was due to a push to be focused purely on numeracy and literacy, as the primary teacher participants had frequently bemoaned. Jessica explained this difference further, saying:

> In the early years it is different …we have the Early Years Learning Framework which is more influenced by the [Federal] Government and sustainability is a big part of that. But I think it's more a push from our curriculum rather than from us. And I know, I've been in my childcare centre for six years, and I've seen when we are going to get assessed so we start implementing things, but then the educators leave and, it's left up to the educators actually implementing it and following it up. Rather than the curriculum stopping us, I think it's us as the educators who cannot keep it going.
>
> *(Jessica, workshop two)*

In this story extract, Jessica suggested it is not 'the curriculum stopping us, I think it's us as the educators who cannot keep it going' as an explanation for limited ESE in early childhood education. She contrasted this against what was happening for the primary school participants who had felt powerless against the school culture and principals pushing them to focus only on NAPLAN testing and results (Barnes, Moore & Almeida, 2018).

From the beginning of the study, Jessica had spoken about the *obstacles* that worked against the implementation of ESE in their long day-care centre. For example, she had spoken of the lack of consistency in staff members, being time poor and minimal persistence in attempting any *environmental education science* programmes. At times, as Jessica suggested in the story extract above, sustainable practices were only put in place on a superficial and temporary basis to appease the Australian Children's Education and Care Quality Authority (ACECQA) assessors. This 'tick the box' attitude for an imminent regulatory assessment of the centre that is revealed in Jessica's stories tends to suggest a deeper resistance to ESE by her colleagues and the Centre as a whole rather than not being *passionate* enough, as Jessica had initially suggested. What is interesting too, in these stories, is that Jessica appears unaware of the hidden pressure she had been under to keep the 'status quo' in line with her colleagues rather than enact her personal and environmental identity to support her professional identity by introducing ESE into the curriculum. While the early career primary teachers in the 'status quo' schools (see Chapter 9) may have felt under pressure to not 'rock the boat' due to school culture pressures, in reality, Jessica was under just as much pressure by the same contextual forces. As suggested in other narrative chapters in this book (see for example, Chapters 10 and 11), the constraints or affordances of a particular context have been shown to be critical determinants in the way early career teachers are enabled or unable to enact ESE in their educational settings.

Despite Jessica's stated confidence in the inclusion of sustainability in the Early Years Learning Framework (EYLF) and the National Quality Framework (NQF), Elliott and Davis (2018) have argued that early childhood teachers 'require guidance to translate EC EfS knowledge into authentic pedagogical practices with children' (p. 173). This guidance is particularly necessary to avoid the anthropocentric approach encouraged through the human-orientated language (such as children's connection with *their* world) embedded within these policy documents (Almeida, 2020). Similarly, guidance on how to integrate sustainability into pedagogical practices has recently become even more necessary, as shortly after our project had finished, ACECQA (2019) removed standard NQF 3.3.1 'to embed sustainability in daily practices' from their nominated requirements. Livingstone (2016) claimed in an online statement that it was 'too challenging' for early childhood educators to achieve this standard, and hence suggested its subsequent removal rather than offer more support to educators to work with this perceived 'challenge' (see Chapter 2). This is an interesting point to make here as Jessica's stories in the next subsection show her difficulties in 'guiding' her colleagues towards ESE in her role as the Educational Leader for the centre

Educational leaders and ESE in early childhood education

Another policy shift in the delivery of early childhood education that impacted on Jessica's personal, professional and environmental identities in her early childhood centre was the introduction of an Educator Leader position. This leadership role had been embedded into the NQF to demonstrate its importance, requiring the Educational Leader to pay 'particular attention to pedagogy… and the responsibility of educational practice…' (ACECQA, Educational Leader Resource, 2019, p. 8), whilst 'inspiring [other educators] to see the possibilities, try new approaches and take professional, calculated risks' (ACECQA, Educational Leader Resource, 2019, p. 16). Within this role, there was a clear expectation that the Educational Leader was to take on the responsibility of 'leading change' as designated 'change agents' within each centre (see Educational Leader Resource, 2019, p. 103).

Despite Jessica's status as an early career teacher, she had been working at the centre in a Diploma of Children's Services position a number of years, and upon graduation, was awarded the role of Educational Leader for the whole centre, as seen here in workshop two:

> I'm [Jessica] and I work at [ ] Early Learning Centre and I'm the Educational Leader. I think lots of people look at me for what to do, especially with programming and what they're doing in their rooms. That's fine with me but I'm more of like everyone has a say and we do it all together, not for me to do it all. So with the environmental educational science I'm interested in it but we don't really do it at the moment…

One of the researchers then asked Jessica if she felt leadership was important for ESE to occur in an early childhood centre, to which Jessica replied:

> Yeah I think so but then I don't really want to make it that you have to do this because then I don't think it'll be done properly, everyone else needs to become passionate as well.
>
> *(Jessica, workshop two)*

However, Jessica's comment 'I don't really want to make it that you have to do this' seems to circumvent the expectation of being a 'change agent' in relation to new pedagogical approaches as the nominated Educational Leader, specifically in relation to EC ESE in their centre. There are a number of aspects to be considered here in the deeper analysis of the societal dimension of Jessica's stories around being the Educational Leader but unable to make any ESE changes in the centre practices. First, there are issues around how the term 'sustainability' is perceived and used by those working in early childhood education, and second, hierarchical power issues that are omnipresent in an early childhood setting, especially in relation to a leadership role. As Uren et al. (2019) have argued, the 'activist' or 'greenie' label is not something well received generally by the Australian public in relation to sustainability, and as a consequence the ESE terminology appears to be avoided. These prevailing attitudes and assumptions appeared to have influenced Jessica's decision not to actively advocate for her colleagues to implement sustainability in their centre, opting instead for a more passive *environmental science education* approach, which she said, 'hopefully everyone else will follow as well….' Later, in workshop five, Jessica again reiterated her hesitancy in actively promoting ESE in their early childhood setting, by saying:

> I can't make them do it, they have to want to do it…But I've been doing it in my room…
>
> *(Jessica, workshop five)*

This suggests that Jessica hesitated to use her position of power as the Educational Leader to shape the field, to dislocate her colleagues' individual habitus in a way that placed ESE in a much more prominent place in their centre. Jessica was also subconsciously furthering the notion that environmental and sustainability issues are deeply held and linked to one's own identities, and that therefore these cannot be imposed on others (Hart, 2003). Further to Jessica's approach to her role as Educational Leader, researchers have found that while the role was established to instigate change in practices, in reality, this rarely occurred (Grarock & Morrissey, 2013; Martin, Nuttall, Henderson & Wood, 2020). For example, Martin et al. (2020) found that 'no national policies are currently in place to specifically support the work of Educational Leaders' (p. 14). Grarock and Morrissey (2013) found in their study of Educational Leaders that 'these teachers were not able to demonstrate leadership to any extent across their centres outside of their own rooms…' (p. 83). Jessica's announcement that she had 'been doing it in my room' supports these findings on a lack of agency in the role of educational leadership beyond the boundaries of her 'own room.' In looking at the temporal and societal dimensions in which Jessica's experience of sustainability is situated, the underlying meanings of these experiences become increasingly visible.

Young children's agency in ESE: *'They get it already'*

The final narrative pattern to discuss in this societal dimension of the analysis of the early career kindergarten teachers relates to children's agentive position as potential change agents in ESE. This is a theme that emerged throughout many of the workshop conversations with many of the early career teachers, the community partners and the researchers across the whole study (see for example, Chapter 11). However, as the review of literature has shown in this early childhood chapter, it is particularly pertinent to focus here on this important theme in relation to young children's agency, voice and transformational power. Returning to the notion of historical informants, Huggins and Evans (2017) have identified the long-held belief of many early childhood educators that children should be protected and sheltered from the 'harsh realities' of life for as 'long as possible' (p. 4). However, in contrast with this attitude, one community partner made a comment to Jessica about the strong presence of young children's agency in relation to the broader dispositions of ESE, to which Jessica readily agreed. The community partner reflected, saying:

> I've had a few kinder groups through the [DRI] centre over the last month and they are all so much more switched on than the primary kids. They get it already, you almost don't have to tell them, they have this natural gravitational pull to injustice and rubbish and picking it up.
>
> *(Community partner, workshop three).*

These comments push back against the dominant discourse around 'innocent' young children needing to be 'protected' against the 'realities of life' (Huggins & Evans, 2017). Instead, the community partner is representing young children as not only agentive in their actions, but in their conscious awareness of what is happening in the world around them (Davis, 2010; Mackay, 2012). Interestingly, in the following workshop four, Jessica's stories started to shift from a concern about her colleagues' dismissive attitude to ESE to a focus on the four-year-old children in her room and their dispositions towards ESE enacted within the playground environment. This was seen, for example, when Jessica spoke about the lunchtime rubbish and the purposeful 'use' of Maggie the Magpie to promote the children's rubbish collection, as seen here:

> And especially having Maggie the Magpie right there in the playground, they're all like, 'There's rubbish, Maggie could eat that', so that really worked for them as well. It's right there…, isn't it?
>
> *(Jessica, workshop four)*

Jessica's tentative steps into talking to the young children in her room about their impact on other-than-human species was evidence of her emerging engagement with ESE and continued into the stories she told in the final workshop five that follow in the next section.

Contextual influences on ESE in early childhood education

The final aspect to discuss in this chapter's three-dimensional narrative inquiry analysis refers to the place-based dimension, and in this instance, this refers to both the early childhood workplace for the early career kindergarten teachers and the notion of place-based learning for the young children in their centres. Both early career kindergarten teachers worked in long day-care settings, with Jessica in a for-profit centre adjacent to the beach some distance away from Melbourne, while Bonnie worked in a community-run, not-for-profit setting under the auspices of the local government in inner-city Melbourne, where she worked as a sustainability educator across a number of centres. Typical of the field, kindergarten teachers can work in a diverse array of places (from family day care to long day care to sessional kindergartens or preschools) with a wide diversity of qualifications (from Certificate III, to Diploma of Children's Services, to Bachelor of Education (Early Years) to a Master of Teaching (Early Childhood or Primary/Early Childhood)) held amongst their colleagues (Molla & Nolan, 2018). As previously noted, the findings identified in this study have shown that the affordances and/or constraints within an educational context are key determinants in how much early career teachers are enabled or not able to enact ESE in their settings. For Jessica, the constraints were primarily based on the workplace structures within the childcare setting where she worked, and for Bonnie, the affordances were enabled through the local government's acknowledgement on the importance of ESE and the EC ESE programme they set up in early childhood education.

A lack of funding and limited on-site support for early childhood ESE programmes was also identified by Jessica's colleagues in their long day-care centre as a financial barrier to *start-up* and continuing any new *environmental science education* programmes. One of the community partners confirmed this predicament by saying:

> My role is to work with Early Childhood Centres on the ResourceSmart program. Although it is funded for schools [through Sustainability Victoria], this funding does not extend to Early Childhood Centres at this stage. As a result, funding is a key issue. In addition, for the program to work educators/teachers need to feel supported. It doesn't work if it is additional work within the same time frame. Getting time release would be another key issue. And finally support from leadership and the community is also essential for a program to be successful.
>
> *(Community partner, ThinkTank workshop)*

The community partner's comments made to the ThinkTank audience, as seen here, are important to note. In one concise statement, she summarised the key issues at hand for any early childhood teacher, especially an early career kindergarten teacher in a long day-care setting, when attempting to enact ESE in their early childhood curriculum. These include not only the lack of funding opportunities for early childhood centres, but also the systemic working conditions of many teachers in childcare settings, where minimal planning time and a lack of community support from time-poor families are commonplace (Kim, Shin, Tsukayama & Park, 2020).

In contrast with this, Bonnie's long day-care scenario was significantly different from Jessica's workplace context. The community partner prompted the following story of a successful implementation of ESE when she asked the community of practice group during workshop two for any *recipes for success* amongst the participants. In response, Bonnie told a story about an ESE project she had initiated first in her early childhood setting, and then broadening out to the whole local government area, when she said:

> I think that projects where we can get our families to be involved with sustainability works well… Like with a story book that I wrote, we have a travelling sustainability bear that goes home with the children, and the children capture ideas where they have been sustainable at home, then come back and do a show and tell about what they did at home with the sustainability bear. The parents come and say, 'we didn't know about this'… We started in one small long day-care centre and now we have a sustainability bear at each centre in [the local government area], with more requests endlessly for the books and bears…
>
> *(Bonnie, workshop two)*

Despite the early childhood field reporting back to ACECQA that sustainability was 'too challenging' to embed into the early childhood curriculum (see Livingstone, 2016, in Chapter 2), Bonnie's story illustrated one way ESE has seamlessly been integrated into the learning experiences already in place in an early childhood programme. In Bonnie's educational context, support from strong proactive leadership was evident in setting up an early childhood sustainability programme, so too Bonnie's sense of confidence to enact her personal, professional and over time, her environmental identity within her workplace.

In contrast, it appeared at the beginning of the workshops that the constraints that were evident in Jessica's workplace were operating against her enactment of ESE in the centre. Upon analysis of Jessica's stories throughout the duration of the study, a shift in her thinking became increasingly visible. She appeared to be more comfortable with her personal and environmental identity, working out ways to use these to reshape her professional identity. It also appeared from her later stories that the collaborative community of practice conversations in our study played a key role in triggering ideas that Jessica subversively started to implement alongside her kindergarten children in their centre's playground and beyond. By the end of the study, Jessica had significantly changed the emphasis of her stories, talking more about how she was starting to use the local beach across the road as the location for her EC ESE outdoor classroom in a meaningful, more targeted way that she had not previously considered, saying:

> ...and we went to the beach today, and next week when we go to the beach, we're going to pick up rubbish. So, we've been starting all of that, which is good. Then when we were at the beach the children were like, "There's rubbish, there's rubbish" and then they all tell me stories of when they're with their parents at the beach and how they're telling their parents to pick up rubbish too. So, it's good...
>
> *(Jessica, workshop five)*

In continuing to use the beach as a meaningful place for the children's localised place-based ESE learning (in a remarkably similar way to Boyd's (2019) study in the UK with the same age group of children and the same outcomes), Jessica broadened her view on the importance of starting ESE in early childhood, where to source recycled materials in her local community and acknowledged the children's agency in influencing their parents' sustainable practices, as seen here:

> [One of the researchers] always talks about the early years and how they should be at the front of all the sustainability initiatives...And it's true... I think that if you build that in at an early age where they go to the beach and they say, 'where's the rubbish?' and then they're telling their parents and it does build this kind of awareness, so it's really important. Yeah, well we've been doing heaps of recycled art, we made balls with scrap paper, we've been getting scrap paper from down the road and I want to try and stop us buying paper except for important things we need to print in the office, so I'm going to ring around and ask companies if they can give us their paper...
>
> *(Jessica, workshop five)*

As evident in Jessica's shifting stories in workshop five, her newly emerging leadership approach towards incorporating ESE into the centre expanded quickly into other small but meaningful sustainable initiatives, such as the intentional use of recycled paper. And then finally, Jessica concluded her stories in workshop five about her increasing engagement with EC ESE by proudly saying:

> So instead of going to the kitchen and the bowls getting washed, after they eat we have a tub of water and they all wash it out, so that's kind of saving water as well, and at the end of the day when we empty our water bottles they all go and take them and put all the water on the trees...So, we do a lot of little things...now.
>
> *(Jessica, workshop five)*

While initially hesitant to use and accept her position of power to promote or force her colleagues to prioritise sustainability, it became clear that Jessica's habitus was increasingly being disrupted, resulting in small yet meaningful changes to her practice and her professional identity. So while Jessica had started with 'we don't really do it,' she had moved towards 'I want to try... I'm going to...' over the course of the project. And finally, at the end of the project, a celebratory acknowledgement that 'we do lots of little things....' As O'Gorman (2020) has insightfully argued, 'transformative teachers need to be bold risk-takers willing to disrupt and rebel against traditional approaches to early childhood education' (p. 190). For Jessica, the first step for her was moving away from the reliance on a collective, institutional habitus that did not prioritise ESE towards taking agentive steps to examine her own ways of doing and thinking.

Looking to the future: hopeful applications

In this chapter, we have outlined the slow but gradual introduction of early childhood ESE into the early childhood field. We have used a three-dimensional narrative inquiry analysis of temporal, societal and place-based dimensions to frame the analytical discussion of the EC ESE literature juxtaposed against two early career kindergarten teachers' lived experiences of ESE in their early childhood settings. As a consequence of this in-depth analysis, we have celebrated one teacher's successful ESE programmes that she has seamlessly incorporated into early childhood programmes within a collective team of 'change ready' teachers, children and families (Morgan, Comber, Freebody & Nixon, 2014) As well, we have identified a significant shift in one of the early childhood teachers' thinking and actions over the duration of the study towards the successful inclusion of *lots of little things* that made up her individual approach to EC ESE within her centre. In doing so, taken-for-granted assumptions about the position of sustainability in early childhood education have been made visible and challenged and as a consequence, has assisted in one teacher's transformative approach to leadership in ESE within an early childhood centre.

References

Almeida, S. C. (2020). Chapter 4: Alternative worldviews on early childhood education for sustainability: Reviewing and re-examining concepts, images of children and sustainability. In S. Elliott, E. Ärlemalm-Hagsér & J. Davis (Eds., pp. 38–50), *Researching early childhood sustainability: Challenging assumptions and orthodoxies*. Abingdon, Oxon: Routledge.

Australian Children's Educational & Care Quality Authority (ACECQA). (2019). *Educational leader resource*. Sydney, Australia: ACECQA.

Australian Children's Educational & Care Quality Authority (ACECQA). *National quality framework* (NQF). Retrieved from https://www.acecqa.gov.au/national-quality-framework

Barnes, M., Moore, D., & Almeida, S. (2018). Sustainability in Australian schools: A cross-curriculum priority. *PROSPECTS*, 1–16. doi:10.1007/s11125-018-9437-x

Boyd, D. (2019). Utilising place-based learning through local contexts to develop agents of change in early childhood education for sustainability. *Education 3–13, 47*(8), 983–997. doi:10.1080/03004279.2018.1551413

Burke, S. (2020). Cited in Duggan, S. Preparing to deal with the emotional toll of this toxic summer. *Australian Teacher Magazine, 16*(1), 16–17.

Chaudhary, M. (2018) Pint-sized powerhouses: A qualitative study of children's role in family decision making. *Young Consumers, 19*(4), 345–357.

Clandinin, J., & Connelly, M. (2000). *Narrative inquiry: Experience and story in qualitative research*. San Francisco, CA: Jossey-Bass Inc.

Cutter-Mackenzie, A., Edwards, S., Moore, D., & Boyd, W. (2014). *Young children's play and environmental education in early childhood education*. Heidelberg, Germany: Springer Briefs in Education.

Cutter-Mackenzie, A., & Rousell, D. (2019). Education for what? Shaping the field of climate change education with children and young people as co-researchers. *Children's Geographies, 17*(1), 90–104. doi:10.1080/14733285.2018.1467556

Dahlberg, G., & Moss, P. (2005). *Ethics and politics in early childhood education*. Abingdon, Oxon: Routledge.

Dahlberg, G., Moss, P., & Pence, A. (2007). *Beyond quality in early childhood education and care: Postmodern perspectives* (2nd ed.). Abingdon, Oxon: Routledge.

Davis, J. (2009). Revealing the research 'hole' of early childhood education for sustainability: A preliminary survey of the literature. *Environmental Education Research, 15*(2), 227–241.

Davis, J. (2010). What is early childhood education for sustainability and why does it matter? In J. Davis (Ed.), *Young children and the environment: Early education for sustainability* (1st ed., pp. 7–31). Port Melbourne, VIC: Cambridge University Press.

Davis, J. (2015). What is early childhood education for sustainability and why does it matter? In J. Davis (Ed.), *Young children and the environment: Early education for sustainability* (2nd ed., pp. 7–31). Port Melbourne, Australia: Cambridge University Press.

Duhn, I. (2012). Making 'place' for ecological sustainability in early childhood education. *Environmental Education Research, 18*(1), 19–29. doi:10.1080/13504622.2011.572162

Early Years Learning Framework for Australia: Belonging, Being and Becoming (EYLF). (2009). Retrieved from https://docs.education.gov.au/documents/belonging-being-becoming-early-years-learning-framework-australia

Edwards, S., Moore, D., & Cutter-Mackenzie, A. (2012).Beyond 'killing, screaming and being scared of insects': Learning and teaching about biodiversity in early childhood education. *Early Childhood Folio, 16*(2), 12–19.

Elliott, S. (2003). *Patches of green: Early childhood environmental education in Australia: Scope, status and direction*. Sydney, NSW: NSW Environmental Protection Authority (EPA) publication.

Elliott, S., Ärlemalm-Hagsér, E., & Davis, J. (2020). Introduction: Reframing the text, a second time. In S. Elliott, E. Ärlemalm-Hagsér & J. Davis (Eds), *Researching early childhood sustainability: Challenging assumptions and orthodoxies* (pp. xx–xxix). Abingdon, Oxon: Routledge.

Elliott, S., & Davis, J. (2018). Chapter 12. Moving forward from the margins: Education for sustainability in Australian early childhood contexts. In Reis, G & Scott, J. (Eds.), *International perspectives on the theory and practice of environmental education: A reader* (pp. 163–178) (Environmental Discourses in Science Education, Volume 3). Switzerland: Springer.

Elliott, S., & McCrea, N. (2016). *Sustainability as a different way of thinking everyday: Examining environmental education in NSW early childhood education services: A literature review with findings from the field*. NSW: University of New England. Retrieved from https://www.environment.nsw.gov.au/resources/grants/160418-Early-Childhood-Report.pdf

Elliott, S., & Young, T. (2016). Nature by default in early childhood education for sustainability. *Australian Journal of Environmental Education, 32*(1), 57–64. doi:10.1017/aee.2015.44

Grarock, M., & Morrissey, A-M. (2013). Teachers' perceptions of their abilities to be educational leaders in Victorian childcare settings. *Australasian Journal of Early Childhood, 38*(2), 4–12.

Hart, P. (2003). *Teachers' thinking in environmental education*. New York, NY: Peter Lang Publishing Inc.

Hirst, N. (2019). Education for sustainability within early childhood studies: Collaboration and inquiry through projects with children. *Education 3–13, 47*(2), 233–246. doi:10.1080/03004279.2018.1430843

Huggins, V., & Evans, D. (2017). Introduction. In V. Huggins & D. Evans (Eds), *Early childhood education and care for sustainability: International perspectives* (pp. 1–12). Abingdon, Oxon: Routledge.

Huntley, R. (2020). Why are we so divided on climate change? *MSSI Oration*. Melbourne University. Retrieved from https://www.abc.net.au/news/2020-01-29/climate-change-global-warming-six-groups-rebecca-huntley/11893384

James, A., Jenks, C., & Pence, A. (1998). *Theorizing childhood*. Cambridge, UK: Polity Press.

Kahn, P. H., Weiss, T., & Harrington, K. (2019). Modelling child-nature interaction in a nature preschool: A proof of concept. *Frontiers in Psychology*. doi:10.3389/fpsyg.2018.00835

Kim, J., Shin, Y., Tsukayama, E., & Park, D. (2020). Stress mindset predicts job turnover among preschool teachers. *Journal of School Psychology, 78*, 13–22.

Lemmon, J. (2020). Personal communication. Stephanie Alexander Kitchen Gardens Early Years Pilot Launch. Clarendon Street Children's Centre, South Melbourne, 29th February, 2020.

Livingstone, R. (2016). ACECQA *We hear you blog: Demystifying sustainability*. ACECQA website blog. Retrieved on August 8th 2016 from https://wehearyou.acecqa.gov.au/2016/08/08/demystifying-sustainability/

Luff, P. (2018). Early childhood education for sustainability: Origins and inspiration in the work of John Dewey. *Education 3–13, 46*(4), 447–455.

Mackay, G. (2012). To know, to decide, to act: The young child's right to participate in action for the environment. *Environmental Education Research, 18*(4), 473–484.

Martin, J., Nuttall, J., Henderson, L., & Wood, E. (2020). Educational leaders and the project of professionalism in early childhood education in Australia. *International Journal of Educational Research*. doi:10.1016/j.ijer.2020.101559

Molla, T., & Nolan, A. (2018): Identifying professional functionings of early childhood educators. *Professional Development in Education*. doi:10.1080/19415257.2018.1449006

Moore, D., Morrissey, A-M., & Robertson, N. (2019). "I feel like I'm getting sad there": Early childhood outdoor playspaces as places for children's wellbeing. *Early Child Development and Care*. doi:10.1080/03004430.2019.1651306

Morgan, A-M., Comber, B., Freebody, P., & Nixon, H. (2014). *Literacy in the middle years: Learning from collaborative classroom research*. Marrickville, NSW: Primary Teachers Association of Australia.

O'Gorman, L. (2020). Chapter 14: Stories of disruption: Perspectives on the use of images to prompt children's action taking for sustainability. In S. Elliott, E. Ärlemalm-Hagsér & J. Davis (Eds.), *Researching early childhood sustainability: Challenging assumptions and orthodoxies* (pp. 179–191). Abingdon, Oxon: Routledge.

OMEP World Congress. (2010). *World organisation for early childhood education.* Retrieved from http://www.327matters.org/Sustainability/Docs/OMEPbroENG.pdf

Platz, D., & Arellano, J. (2011). Time tested early childhood theories and practices. *Education, 132*(1), 54–63.

Siraj-Blatchford, J., & Huggins, V. (2015). Thinking global and acting local, and thinking local and acting global: Sustainable development in early childhood care and education. *Early Education Journal, British Association for Early Childhood Education,* No. 76. Retrieved from http://www.schemaplay.com/Docs/TGAL.pdf

Stephanie Alexander Kitchen Gardens (SAKG) Early Years Pilot Launch. (2020). *Good Food* – Planting healthy habits: Stephanie Alexander Kitchen Garden rolls out to preschoolers. Retrieved from https://www.goodfood.com.au/recipes/news/planting-healthy-habits-stephanie-alexanders-kitchen-garden-rolls-out-to-preschoolers-20200227-h1m473

Stuhmcke, S. M. (2012). *Children as change agents for sustainability: An action research case study in a kindergarten.* Unpublished thesis for Doctor of Education degree. Brisbane, QLD: Queensland University of Technology.

Thunberg, G. (2019). *No one is too small to make a difference.* London, UK: Penguin Press.

United Nations Declaration on the Rights of the Child. (1989). *Office of the United Nations high commissioner for human rights,* Article 12. Geneva, Switzerland: United Nations, Committee on the Rights of the Child Publication.

United Nations Educational, Scientific, and Cultural Organisation. (2010). *Climate change education for sustainable development.* Retrieved from https://unesdoc.unesco.org/ark:/48223/pf0000190101

Uren, H., Dzidic, P., Roberts, L., Leviston, Z., & Bishop, B. (2019). Green-tinted glasses: How do pro-environmental citizens conceptualize environmental sustainability? *Environmental Communication, 13*(3), 395–411. doi:10.1080/17524032.2017.1397042

Wals, A. E. J., & Benavot, A. (2017). Can we meet the sustainability challenges? *European Journal of Education, 52*(4), 404–413. doi:10.1111/ejed.12250

9

ESE IN STATUS QUO SCHOOLS

'It's just not a priority'

Box 9.1 Extract from Kath in workshop three

I promise I'll try not to cry but I'm in a pretty vulnerable state at the moment because I'm really not enjoying my job. But there's the learners who are considered at risk but they're six months behind level.... These are the learners who are getting taken out of every other subject for integration support for reading, writing and maths. They're being taken out of subjects like our integrated unit where we're looking at community, they're being taken out of science, they're being taken out of health, out of Bounce Back which is a resilience programme. They're being taken out of these classes just to focus on reading, writing and maths outcomes. So, it's just devastating.

(Kath, suburban school, workshop three)

Introduction

Over the past decade there has been a global increase in *national* education policy reforms, which aim to provide national consistency in schools and increase a country's international competitiveness within the global education market (Barnes & Cross, 2018; Savage, 2016). In Australia, in particular, increased national control over education has resulted in reforms such as a national curriculum, standardised national literacy and numeracy assessments, professional standards for teachers and school leaders and a national school funding model (Savage, 2016). These national reforms have resulted in not only increased national consistency but increased accountability within Australian educational contexts. Standardised literacy and numeracy tests have been introduced in Australia (testing all learners in Years 3, 5, 7 and 9) and the results of these tests have been published on the federal government's *My School* website. The rationale behind the website when introduced in 2010 was to provide information to parents so they could (1) select a school for their children and (2) put pressure on principals to improve their schools' test results (Mills, 2015). This, therefore, has resulted in an imbalanced focus on literacy and numeracy skills—limited, however, to the specific skills assessed on the test—and increased pressure on schools to teach to the test (Fehring & Nyland, 2012). Literacy and numeracy skills have become proxies for student success as well as identify a 'good' school from a 'bad' one. Therefore, what impetus is there to prioritise Environmental and Sustainability Education (ESE) in schools if it is not a viable measure of what determines a 'good' school?

This chapter explores the stories of four early career teachers as they articulate their struggles in understanding what should be taught in order to provide a 'good' education, particularly given their personal and environmental identities regarding the role of ESE in schools. This chapter begins by examining how Australia is positioned within the international education race that, over the last decade, has introduced a national assessment system, national curriculum and teacher and principal standards in order to provide consistency and improve educational outcomes across Australia. This, however, raises concerns about teacher agency in an era of standardised testing and accountability. Situated within these current policy trends, our three-dimensional narrative inquiry analysis (Clandinin & Connelly, 2000) unpacks these four early career teachers' stories, revealing the tensions that exist when early career teachers attempt to implement ESE within a school that does not see it as a priority. This analysis highlights how these early career teachers negotiate institutional norms in their first few years of teaching (temporal), second, their perceptions on how their school prioritises ESE (societal) and third, their views on the available spaces and places for ESE that remain after 'core' learning occurs (place-based).

Understanding how Australia is positioned within the global education race

With the increasing awareness that a nation's educational outcomes have a direct impact on where nations sit relative to one another in an era of 'informational capitalism' (Ball, 2013, p. 1), education policy reforms have been initiated to help keep pace in the global education race. In order to assess where nations are positioned within this education race, international assessments such as the Programme for International Learner Assessment (PISA) have focused on the skills of literacy, numeracy and science to judge the effectiveness of a nation's education system. Scores on the PISA, which tests 15-year olds in 72 countries around the world, play an important role in deciding where nations rank amongst one another (Barnes & Cross, 2018). Yielding a great deal of power, the decline in PISA scores over the past several years in Australia has resulted in a suite of educational reform measures to improve the quality of education (Barnes & Cross,

2018; Baroutsis & Lingard, 2016). In other words, these international bench-marks have harnessed great power in directing education policies. International assessments, such as PISA, have not only created a culture of performativity amongst nations, they have influenced how countries, such as Australia, rely on standards and standardised tests within their own education systems to determine what particular skills are important for a strong education system.

Australia's National Assessment Program – Literacy and Numeracy (NA-PLAN) test was introduced in 2008 and was followed by the creation of the *My School* website in 2010, which aimed to provide both transparency and accounta-bility within Australian schools by publishing NAPLAN data for school compar-isons (Fehring & Nyland, 2012; Mills, 2015). There have been concerns raised about the unintended consequences of NAPLAN, including a narrowing of the curriculum and teaching to the test (Fehring & Nyland, 2012; Johnston, 2017). Rose et al. (2018) conducted a systematic review to determine the dominant discourses surrounding NAPLAN and identified the prevalence of the following four overarching discourses within the literature: 'datafication,' 'social justice,' 'affect and emotion' and 'accountability and performativity.' Their review rec-ognised the benefits of NAPLAN, including the identification of key life skills, targeting funding for schools with the greatest amount of need and bringing awareness of the 'scandalous gap' between the educational outcomes of Indige-nous and non-Indigenous learners (Rose et al., 2018, p. 12). However, they also argued that for standardised testing to be beneficial, it needs to provide substan-tial information on both learning strengths and gaps as well as ensuring that teachers and learners are well informed and believe in the purposes of the test.

In most cases, however, there seems to be a focus on the gaps rather than the strengths. Tests, such as NAPLAN, are used to identify academic achievement gaps which usually result in pressure being placed on schools—and teachers—to close that gap (Miranda, Radcliff & Flora, 2018). While identifying and address-ing academic achievement gaps is an important role that teachers and schools must play, the increased national control over what is taught in Australian class-rooms, through national tests and a national curriculum, has left many educators feeling a sense of disempowerment as they struggle for greater autonomy in their role as educators.

Teacher agency in an era of standardised testing

In an attempt to improve overall performance in schools at a national level, the government has often employed strategies that increase control over teachers' work (Kostogriz & Doecke, 2013). In Australia, an increase of national control has been realised over the past decade with the establishment of the NAPLAN, a national curriculum and national teacher and principal standards (Savage, 2016). The push for accountability and standardisation across Australian schools has left many educators feeling disempowered. Sahlberg (2010) argues that stand-ardised testing has left teachers feeling 'trapped,' as they desire to teach with a moral purpose and build social capital within schools but must be accountable to a competitive, efficiency-driven environment that focuses primarily on measurable achievement (p. 49). If educational success and education quality can be effectively and solely measured in numerical terms, we dismiss vital aspects of the teaching and learning process. Kostogriz and Doecke (2013) attempt to capture the complexities of the classroom, which cannot be understood simply by a world of numbers:

> The world of numbers—the world of *My School*—is a place of abstraction in which particular teaching practices and events, as well as social relationships between teachers and learners, disappear. The world is indifferent to the everyday life of teachers in schools, in their unique locations in communities, their decision-making about what and how to teach, and their situated responsibility for learners.
>
> *(p. 91)*

Marginalised and/or underperforming schools often become the targets of education reform initiatives that seek to improve test scores rather than respond to local needs. Colegrove and Zuniga (2018) suggest that there is a tendency for underperforming primary schools to focus on highly structured, remedial learning programmes to satisfy national and state standardised testing regimes rather than provide rich learning experiences that embrace the community's uniqueness. This preoccupation with keeping pace in the education race has stripped away teachers' freedom and agency to respond to the unique needs of their learners and their communities. Early career teachers, in particular, are positioned precariously as they attempt to develop their professional identity within an era of standardisation and control. In the context of Australia, increasing national control, which has prioritised literacy and numeracy skills, has left many early career teachers with a limited sense of agency to provide rich learning experiences that allow them to embrace their ecological identities.

The teachers' stories: a three-dimensional analysis

Four of the 11 early career teachers in this study taught at schools that we have characterised as being a 'status quo' school. In other words, the priority of the school was to ensure that learners were performing well, particularly in comparison to other schools, and providing remedial learning for the skills deemed most important. This narrative inquiry analysis investigates the temporal, societal and place-based dimensions of these teachers' stories to illustrate the tensions that exist when early career teachers attempt to implement ESE in their classrooms. The four teachers (Mary, Kath, Daniel and Rick) were all teaching at the primary level, with two schools having just completed a school review which put a great deal of pressure on these schools to improve learner performance on the NAPLAN. First, the three-dimensional analysis reveals the tensions that exist when early career teachers attempt to understand and negotiate institutional norms within their first few years of teaching and fuse their professional and ecological identities during this time. Second, the analysis exposes the societal pressures that arise when schools must perform and therefore prioritise what learning counts. Finally, the places and spaces for ESE, both physically and in the curriculum, are discussed to expose how ESE is prioritised in these four schools.

Developing as an early career teacher: a time-consuming process

Many empirical studies in education report that teachers often claim that they lack the time to adequately meet the needs of their learners due to the time required for planning, providing additional support for some learners and fitting in all aspects of the curriculum (Barnes, Shwayli & Matthews, 2019; Caldwell, 2015;). However, the time-consuming nature of meeting learner needs is intensified for early career teachers, as they must not only manage their own time but also negotiate the institutional norms, practices and priorities of their school. Many scholars suggest that early career teachers' instructional practices are shaped by their school contexts over their first years of teaching (Vagi, Pivovarova & Miedel Barnard, 2019). The institutional or collective habitus of their school therefore reshapes their professional identities and their professional practices, further validating their personal and environmental identities.

In the case of our four teachers, the negotiation of institutional norms first occurs when they realise that there is a clash between their ways of thinking, or their habitus, and that of the school.

> It's how you juggle what you want to do versus what you need to do versus what you're told to do.
>
> *(Daniel, workshop four)*

> I think that may be a bit of a frustration when you first come out of uni[versity] and have all these ideas of what - we know what we should be doing. We know what we're passionate about. We've seen the research. We know it. Then when you get out there, it's like, but I can't do that? That's not important?
>
> *(Kath, workshop four)*

While both Daniel and Kath experienced what they felt they should do, based on their identities, their passions, convictions and prior experiences at university, and what they can or are told to do, were sometimes at odds with one another. This created a disruption to their habitus or their current ways of thinking and doing and required time spent on negotiating their habitus with the school's collective habitus. It led to a dissonance between their personal and environmental identities and how their professional identities were being shaped. However, through the process of our workshops, Kath articulated that it takes time and patience for others to understand and realise one's passion for ESE. The desire and passion to care for the environment and value the connections between the environment and one another is not something that can be transmitted passively to others. Instead, it is a process that requires others to take ownership over time:

> …absolutely it takes time and patience. I think those slow steps will help develop and build other people to invest their own time into it and their own ownership and to connect with them and they'll be more inclined to be a part of it…. So just taking those really slow steps is one of the big things I took from last session…
>
> *(Kath, workshop four)*

These slow steps refer to the multitude of questions, discussions, activities and instructional practices that teachers can take to ignite a passion within learners, colleagues and school leaders, and allow them to invest personally. This revelation holds great importance as Kath comes to the realisation that while her convictions and passion for ESE are vital to her overall identity, those around her in her school setting must be given the time to connect to ESE in a way that is meaningful for them.

In addition to coming to terms with their evolving professional identities, which embraces their environmental identities in new and different ways, the teachers found that prioritising ESE was difficult in an already crowded curriculum:

> I did find at the end of the term with assessments and everything it was very difficult to fit it all in… That's what I've found this year and it has been really hard, actually. You might have some ideas and they're happy for you to take the children outside and that sort of thing, but you've still got to tick everything off that week and what you need to cover.
>
> *(Mary, workshop four)*

> …they were talking about planning out the next 12 months, and then one of the head teachers came in and said, 'These are all the focuses, and everyone wants their time, something's got to give.
>
> *(Daniel, workshop five)*

Mary acknowledges that in her school, she has some freedom to implement ESE in the classroom; but there is the difficult task of attempting to cover everything while also weaving sustainability themes into the many topics she must cover. Similarly, Daniel highlights that in the juggle to cover everything, usually something has to give and, in many cases, this tends to be ESE, as discovered in our previous research (Barnes et al., 2019). Kirkby, Moss and Godinho (2017) argue that early career teachers are often expected to manage a workload that is similar to their more experienced colleagues and to navigate their evolving professional identities on their own. Consequently, early career teachers' professional identities are being shaped by the educational field and/or the collective habitus of their school, yet all the while they are not adequately supported in negotiating their workload, in attempting to cover everything in the curriculum or mentored in navigating their evolving professional and environmental identities.

Is ESE a priority in Australian schools?

Given the time constraints to cover as much of the curriculum as possible, which requires prioritising particular knowledge and skills over others, the participating teachers from these status quo schools reported that they felt that, in their particular schools, ESE often did not make the priority list. With debates surrounding the overcrowded Australian Curriculum in primary and secondary schools, Marsh, Clarke and Pittaway (2014) suggest that this is due to the multiple and diverse education stakeholders who all have different ideas about what should be prioritised in schools. Therefore, they argue, these different priorities should be carefully considered and appropriately contested, where needed. However, as discussed above, many early career teachers struggle to navigate the tensions between the seeming priorities of the school with their own convictions about ESE. Several of the teachers explicitly stated that ESE was not a priority and they found it difficult to receive support in implementing sustainability themes within their classroom practices:

> So, I'm finding the school that I'm at, it is not a priority whatsoever and that is a big challenge that I'm facing to even implement little things in my classroom.
>
> *(Kath, workshop two)*

> Their priority isn't [ESE] - it's fitting those [other curriculum areas] things in.
>
> *(Mary, workshop four)*

> It's just hard, you don't get support from – not that my school's not supportive, but it's just not a priority, it's not an interest.
>
> *(Rick, workshop two)*

Unlike Kath, who felt that she had little support to implement ESE in the classroom, Rick found that his school is generally supportive, yet they still did not prioritise ESE. Interestingly, he mentions that it is not of interest. In contrast to Chapter 7, where the participating early career teachers felt that their school implemented a whole-school approach to ESE because it was part of their ethos or collective habitus as a school, Rick's school did not share that same ethos. Mary suggests that ESE is not a priority because it must compete with more pressing curriculum areas. This may suggest that national initiatives like the NAPLAN and My School place pressure on schools to prioritise skills such as literacy and numeracy in an overcrowded curriculum. In other words, this exposes a knowledge hierarchy that is supported by societal agendas (Miller & Windle, 2010). Kath's comments about the NAPLAN expose this knowledge hierarchy:

> ...they [her school] did go through a peer review and their focus is all about improving maths, reading and writing targets. That is their focus and it's all about NAPLAN... Last year it was reading, year before it was maths, now it's writing. They want to see data, improvements in targets and it's [ESE] unfortunately just not a priority... I don't understand why they're just picking and choosing that part [reading, writing and maths] and not this part [ESE] as well.
>
> *(Kath, workshop three)*

Kath's comments reveal the frustration that comes with being an early career teacher who is attempting to understand the knowledge hierarchy in schools—or more specifically understanding the field of education—what counts and what does not. Importantly, Kath, Rick and Daniel comment on their schools' desire for improvement and how this desire often marginalises ESE as a knowledge area:

> ...we just went through a review two years ago so it's [ESE] not the priority.
>
> *(Kath, workshop two)*

> Our school went through a review not last year but the year before. Environment or sustainability is not really a priority even though they sell it on the curriculum, and I take it on board, or try to do it myself and then seek permission later, or forgiveness basically... The politics in the school, or working for the government anyway [teachers in public schools are government employees], is just ridiculous.
>
> *(Rick, workshop two)*

> There's a lot of school improvement stuff going on at our school at the moment, so we're doing big workshops on evidence on how we're filling certain criteria... principals are competitive against other schools in their area as well. Because they're competing for the same learners as well. So, they're making sure that they're doing everything that they need to do to be able to be attractive...
>
> *(Daniel, workshop four)*

These comments suggest that there is a knowledge hierarchy and this hierarchy is established by meeting particular criteria, mainly through improving performance on the NAPLAN. Given that ESE is not a form of content that is assessed on the NAPLAN, it is easily brushed aside in a bid to be competitive in the

education race and be attractive in the educational field. However, the participants suggest that while ESE is not prioritised, it is still something they have to
tick off the list, often in a tokenistic manner, as shown by Rick's comment above.
Kath and Daniel report how their school attempts to tick ESE off the list by positioning it in mission statements and in professional staff focus groups:

> The mission statement…So, I suppose, on paper, it looks like, oh yeah,
> they're doing all of that. But if you actually go into the school, it's a bit
> tokenistic at times, I suppose.
>
> *(Kath, workshop four)*

> … it was getting crowded for the planning for next year, and there was even
> talk about, "Do we even have sustainability as a focus community?" …
> we'll just tick that box, it'll come together, and life will be good… It [ESE]
> was on the outer.
>
> *(Daniel, workshop five)*

Overall, the participating teachers felt that ESE was identified as a curriculum
area that needed to be addressed and/or checked off, but it did not hold any
particular status or importance within their schools. When compared with the
knowledge and skills that could quantify the school's performance against other
schools (e.g. literacy and numeracy skills), ESE had limited pull in shaping the
school ethos.

Finding a place for ESE in status quo schools

Given the priority placed on literacy and numeracy skills in order to keep pace
with other schools, one of the schools represented in this chapter withdrew learners from the classroom during 'integrated subjects' which would allow for the
integration of ESE themes. This disruption to the classroom resulted in some
learners missing out and made it difficult to organise meaningful place-based
opportunities that built a sense of community action within the classroom, as the
teacher did not want to leave people out. The following exchange, between Kath
and one of the community partners, contextualises the quote used at the beginning of the chapter and illustrates the early career teacher's struggle to make ESE
meaningful when learners need to show that they are progressing in reading,
writing and math outcomes:

> …there are the learners who are considered at risk but they're six months
> behind level… These are the learners who are getting taken out of every
> other subject for integration support for reading, writing and maths.
> They're being taken out of subjects like our integrated unit where we're
> looking at community, they're being taken out of science, they're being
> taken out of health, out of Bounce Back which is a resilience program.

They're being taken out of these classes just to focus on reading, writing and maths outcomes.

(Kath, workshop three)

They're trying to marry the modern ideas with the 1950s rote learning, and as teachers we just have to recognise that's the way it is and sometimes you can win a battle and sometimes you can't… Particularly for you [Kath]. Your biggest issue is some of these kids are taken out, how is that affecting you and your classroom learning other than the kids themselves are missing out on those units?

(community partner, workshop three)

Well yes, they all come and go and then they come back into the classroom and then it's like, 'What have we just done, what are we doing?" and you have to catch them back up on that. Then they start feeling like I don't know what's happening and so then they become disengaged because I've missed that whole chunk, it doesn't matter so then they're missing that… And then they fall behind even more.

(Kath, workshop three)

This exchange brings to light several tensions at play in current classrooms. The first, as discussed earlier, is that it can be difficult to build a sense of community and continuity when learners are being taken out at various points within the day. The community partner suggests that not only does the withdrawal of learners during class time influence those learners, but also the classroom's collective learning. Classroom transitions provide timely challenges for learner disengagement and require teachers to have strong classroom management skills (McCurdy et al., 2018). As an early career teacher, it can be difficult to manage the classroom and keep a natural flow and energy within classroom activities when there are constant disruptions, such as learners packing away their school supplies to attend a remedial lesson. This speaks to an early career teacher's sense of control and agency as they must negotiate the school's ways of doing, or habitus, and attempt to respond in a way that allows them to still feel as if they have control over their classroom. It also speaks of the constant battles to address conflicts between their personal and environmental identities with their professional identities as teachers. The community partner suggests you can win some battles and others you cannot. Therefore, Kath must come to terms with what she has control over and what she does not, while also devising a plan of action so she can ensure the class is collectively learning and progressing.

Second, learners begin to identify that integrated units, or units that allow for the integration of topics such as ESE, are not a priority and therefore 'catching up' on these topics is considered not as important as other topics. Given that many scholars suggest that taking learners out of mainstream activities can deprive them of important sources of knowledge (Zyngier, Black, Brubaker & Pruyn,

2016), it becomes important to consider how remedial programmes position other content areas and how they can promote disengagement with topics that might provide relevant learning opportunities for these underperforming learners. Drawing on Bourdieu's concept of social reproduction, underperforming learners become trapped in a holding pattern in which they try to catch up, but in exchange for something else. These learners fall into a pattern of inequality that sees them progressively falling behind their peers. Zyngier et al. (2016) suggest that remedial and alternative programmes must be a last resort for schools, with prevention and early intervention as the first step. Therefore, schools need to be proactive in tackling underperformance rather than responding with remedial programmes to address their school's decline in NAPLAN scores.

Despite Kath's experiences with learners being physically withdrawn from the classroom, influencing the classroom dynamics and testing her sense of control and agency, several of the teachers found agentic ways to integrate ESE by utilising outdoor spaces:

> We do have a small garden and they have a garden club on Monday which I'm now involved in, but it's just for Year 2 to 6 so a lot of the kids haven't been to the garden before... So, I got to take each grade out one lesson each and we were exploring nature and took them to the garden, then we came back and talked about what we saw... Then we had another session where we were talking about looking after our environment and recycling and that type of thing.
>
> *(Mary, workshop two)*

> I teach PE, but then how do I incorporate from a cross-curricula point of view?... how do I get these kids forming a connection with place and love for the outdoors through PE? That's something that's stayed in the back of my mind. We do our PE lessons - we've got a big oval, we've got basketball courts, all that built stuff. But we've got this amazing outdoor play area that got rebuilt last year that's in the trees. There's a lot of nature in there. So, we do a lot of our PE, either warmups or actual games, through there. I've actually found that kids are starting to pick up rubbish when they go through there now, which they never used to do. They used to just walk straight past it... Because it's a place that they enjoy. So that's something that I can do within the confines of the school...
>
> *(Daniel, workshop four)*

For both Mary and Daniel, they had to identify what they could do 'within the confines of the school.' For Mary, it meant she joined the garden club, which many learners had never visited, and volunteered to take out each grade outside for one lesson. In contrast to Kath, while Mary's school may not have prioritised ESE, they allowed her the autonomy to make connections to ESE, utilising the *existing* school garden. Similarly, Daniel found a way to identify the spaces that currently existed within the school to allow for meaningful ESE connection. Given that there was an outdoor playing area that provided a nature strip of trees, he (re)envisioned how he might use this space to link physical activity with environmental sustainability. Building on the tenets of place-based education (see Chapter 7), Daniel was able to provide an opportunity for his learners to make a meaningful connection between physical education and nature and take ownership of a place that they enjoyed.

Looking to the future—hopeful applications

The classroom can at times be a site of struggle for early career teachers as they respond to the tensions between their professional and ecological identities, while also trying to understand their schools' collective habitus or ways of being, so they can become legitimate members of their new school community. However, this chapter attempts to highlight that these teachers are agentic, and they respond, contest and modify their beliefs and practices in a way that seeks to accommodate their schools' habitus while still recognising their own habitus, personal convictions and identities.

While Mary and Daniel found solace in the spaces available in their schools that allowed them to connect to ESE into their own classes, Daniel's participation in this study and his involvement in our community of practice promoted agentic actions to embed ESE through a cross-curricular and whole-school approach:

> I think last time we met we were looking at potentially having focus groups in our school so that we got little communities within the school, and I was put onto STEAM [Science, Technology, Engineering, Arts and Mathematics]. So we had a STEAM meeting and they were talking about planning out the next 12 months, and then one of the head teachers came in and said, 'These are all the focuses and everyone wants their time, something's got to give' **and so I took your ideas**, I said, 'Why don't we put sustainability in with STEAM' and they were like, 'Right, how would that work?' and we talked about these different projects that we're doing and they're like, 'Right, one of them is going to be sustainability and STEAM together.' So, a little bit excited by that. But it made me think that, at my school at least, sustainability needs to be integrated as that cross-curricular thing, it doesn't work for us as a standalone, it needs to be joined in with other things. So, I was pretty excited about that, I wanted to share that.
>
> *(Daniel, workshop five)*

Whether articulated or not in this extract, Daniel had a sense of relief and joy that as an early career teacher he was able to proactively find a way to embed and integrate ESE themes within current curricular projects. In addressing the temporal, change can occur over time as teachers actively negotiate their roles as early career teachers, but also embrace their agency as they find physical spaces and/or curriculum connections to further ESE ideas within a school that may acknowledge its importance but struggles to prioritise it.

References

Ball, S. (2013). *The education debate* (2nd ed.). Bristol, UK: Policy Press.

Barnes, M., & Cross, R. (2018) 'Quality' at a cost: The politics of teacher education policy in Australia. *Critical Studies in Education*. doi:10.1080/17508487.2018.1558410

Barnes, M., Shwayli, S., & Matthews, P. (2019). Supporting EAL learners in regional education contexts. *TESOL in Context, 28*(1), 45–63.

Baroutsis, A., & Lingard, B. (2016). Counting and comparing school performance. *Journal of Education Policy, 32*(4), 432–449. doi:10.1080/02680939.2016.1252856

Caldwell, B. (2015). Feeling overwhelmed? It is time for serious innovation. *Australian Educational Leader, 37*(1), 14–17.

Clandinin, D. J. & Connelly, J. (2000). *Narrative inquiry: Experience and story in qualitative research*. San Francisco, CA: Jossey-Bass Inc.

Colegrove, K., & Zuniga, C. (2018). Finding and enacting agency: An elementary ESL teacher's perception of teaching and learning in the era of standardised testing. *International Multilingual Research Journal, 12*(3), 188–202.

Fehring, H., & Nyland, B. (2012). Curriculum directions in Australia. *Literacy Learning: The Middle Years, 20*(2), 7–16.

Johnston, J. (2017). Australian NAPLAN testing: In what ways is this a 'wicked' problem? *Improving Schools, 20*(1), 134. doi:10.1177/1365480216673170

Kirkby, J., Moss, J., & Godinho, S. (2017). The devil is in the detail: Bourdieu and teachers' early career learning. *International Journal of Mentoring and Coaching in Education, 6*(1), 19–33.

Kostogriz, A., & Doecke, B. (2013). The ethical practice of teaching literacy. *Australian Journal of Language and Literacy, 36*(2), 90–98.

Marsh, C., Clarke, M. and Pittaway, S. (2014). *Marsh's becoming a teacher* (6th ed.). Frenchs Forest, NSW: Pearson Australia.

McCurdy, M., Skinner, C., Ignacio, P., & VanMaaren, V. (2018). The timely transitions game. In R. Hawkins & L. Nabors (Eds.), *Psychology of emotions, motivations and actions. Promoting prosocial behaviors in children through games and play* (pp. 29–61). New York, NY: Nova Science Publishers.

Miller, J., & Windle, J. (2010). Second language literacy: Putting high needs ESL learners in the frame. *English in Australia, 45*(3), 31–40.

Mills, C. (2015). Implications of the My School website for disadvantaged communities: A Bourdieuian analysis. *Educational Philosophy and Theory, 47*, 146–158. doi:10.1080/00131857.2013.793927.

Miranda, A., Radcliff, K., & Flora, O. (2018). Small steps make meaningful change in transforming urban schools. *Psychology in Schools, 55*(1), 8–19. doi:10.1002/pits.22094

Rose, J., Low-Choy, S., Singh P., & Vasco, D. (2018): NAPLAN discourses: A systematic review after the first decade. *Discourse: Studies in the Cultural Politics of Education*. doi:10.1080/01596306.2018.1557111

Sahlberg, P. (2010). Rethinking accountability in a knowledge society. *Journal of Educational Change, 11*, 45–61. doi:10.1007/s10833-008-9098-2

Savage, G. (2016). Who's steering the ship? National curriculum reform and the reshaping of Australian federalism. *Journal of Education Policy, 31*(6), 833–850. doi:10.1080/02680939.2016.1202452

Vagi, R., Pivovarova, M., & Miedel Barnard, W. (2019). Keeping our best? A survival analysis examining a measure of preservice teacher quality and teacher attrition. *Journal of Teacher Education, 70*(2), 115–127. doi:10.1177/0022487117725025

Zyngier, D., Black, R., Brubaker, N., & Pruyn, M. (2016). Stickability, transformability and transmittability. *International Journal of Child, Youth and Family Studies, 7*(2), 178–197. doi:10.18357/ijcyfs72201615717

10

THE HIERARCHICAL SCHOOL AND ESE

'New teachers cannot do anything'

Box 10.1 Extract from community partner in workshop two

My advice would be that you always take baby steps and you find out where the hierarchies are … And learn that you have to play the game a little bit….

(Community partner, workshop two)

Introduction

The transition from a pre-service teacher to an early career teacher is often typi-fied as being a turbulent and lengthy journey, as these teachers 'grapple with the multitude of tasks, duties, and responsibilities associated with fulfilling their new roles as "real" teachers' (Johnson et al., 2014, p. 530). With worries about high teacher attrition rates (Kraft & Papay, 2014; Weldon, 2018), many scholars have attempted to explore the experiences of early career teachers during these years and consider ways forward to best support them (Arnup & Bowles, 2016; Burke, Aubusson, Schuck, Buchanan & Prescott, 2015; Byrne, Rietdijk & Pickett, 2018; Hume, 2013; Weldon, 2018).

A recent newspaper article from *The Age*, a Melbourne-based newspaper, re-ports the introduction of a new mentoring programme for graduate teachers in the state of Victoria (Carey, 2020). This state mentoring programme, which is to be piloted with 700 teachers in 2021, seeks to address the problems associated with retaining young teachers in the profession. Quoting the Australian Educa-tion Union's Victorian branch secretary, Meredith Peace, the author argues 'that the first couple years of teaching are the most challenging and that burnout is more prevalent amongst young teachers than ever before' (Carey, 2020, para. 10). The article laments that many new teachers equate their experience with the act of being thrown off a bus and they must learn to navigate these tricky (and painful) years of teaching. This chapter explores how the participating teachers in this study, with a specific focus on one in particular, attempt to understand their roles as early career teachers and persevere through these unsettling years. As one of the community partners explained in workshop two of this study (as seen above), early career teachers must be strategic in how they 'play the game' in their first years of teaching, particularly if they want to implement ESE ideas and practices in their schools. She explains that teachers must first understand the hierarchies that govern school contexts, ensure that they receive the appropriate backing and support of principals and/or school leaders and take small yet intentional steps forward.

This chapter re-narrates the stories of one early career teacher, Waruni, whose desire to share her Environmental and Sustainability Education (ESE) ideas and practices with her colleagues resulted in feelings of disappointment and regret. This chapter begins by exploring how early career teachers are positioned within schools, often depicted by what they lack rather than what they have to offer. This positioning is examined with regard to how it contributes to high teacher attrition rates and the role of school contexts in developing early career teachers. Following from this brief review of literature, the three-dimensional narrative inquiry analysis (Clandinin & Connelly, 2000) retells the stories of Waruni and examines her experiences in a school that is characterised as being 'hierarchical' in nature. The analysis revealed that creating change not only takes time through intentional small steps (temporal), but that it is a journey that requires awareness of the social context of schools (societal). This analysis highlights the importance of early career teachers having a space to share ideas and practices with like-minded people (placed-based) and how these spaces allow for these teachers to feel legitimised in their convictions and help them persevere through the struggles associated with the first few years of teaching. It showcases how opportunities for developing, strengthening and reshaping teachers' identities are critical, especially for early career teachers.

Early career teachers: is it what they lack or what they bring to their school communities?

There has been an increased focus, in national and state education policies around the world, to attract, select and retain high-quality teachers (Australian Government Department of Education and Training, 2016; Australian Council for Education Research, 2020; Dover, 2018). However, Kraft and Papay (2014) argue that the context, or the professional environment, plays a role in not only developing effective teachers but maintaining them. Specifically, they argue that when school leadership provides opportunities to collaborate and recognise teachers for their efforts, this has the potential to promote teacher improvement. This suggests that supportive environments that enhance individual teachers' personal and professional identities and allow for stronger confluence can ultimately lead to teacher empowerment.

However, early career teachers are often positioned as 'novices,' especially given that the first years of teaching are marked with quick gains of improvement in contrast to later in their careers where improvement appears to be more modest (Boyd, Grossman, Lankford, Loeb & Wyckoff, 2009; Kraft & Papay, 2014). Even the term 'novice' positions these teachers at a deficit in comparison with their more experienced peers and it can easily shift the focus on areas for improvement rather than gains. While limited, there are a number of studies that highlight the ways in which early career teachers can lead others, bringing with them evidence-based knowledge and approaches and their own personal experiences (Byrne et al., 2018; Hume, 2013). For example, Hume (2013) found that early career teachers were able to lead others in their schools as they brought with them innovative teaching approaches and new ways of interrogating the primary science curriculum. Therefore, while the early years of a teacher's career may be marked by rapid gains in overall improvement, this does not mean they are precluded from being able to provide new ideas and/or lead others.

Given the anticipated rate of improvement during this early career period, schools are situated to not only provide supportive professional environments that assist early career teachers' development but also that develop their leadership skills. How the school views their potential to lead in the areas that they excel and/or are passionate about in line with their personal identities has an impact on their evolving professional identities. Early career teachers have the potential to thrive in environments where there is mutual trust and respect (Bryk, Sebring, Allensworth, Luppescu & Easton, 2010). Principals, in particular, play a key role in supporting teachers not only in their development but in becoming leaders. Their ability to effectively support and recognise teachers' effective practices has an incredible influence on a teacher's decision to remain at their school. Early career teachers need to be recognised for what they bring to the community, not just what they lack.

Battling high teacher attrition rates—Supporting yet empowering teachers

Concerns regarding a high teacher attrition rate are well documented in research (Kraft & Papay, 2014; Weldon, 2018). Weldon (2018) reports that while the media, journal articles and research have estimated Australia's early career (first five years) teacher attrition rate at 30%–50%, these estimates are not well established, but still highlight that teacher attrition is a wicked problem that requires attention. Many scholars attribute this problem to the lack of support provided to early career teachers during the first five years of their careers, with the need for comprehensive mentoring and collegial support (Arnup & Bowles, 2016; Weldon, 2018). While there has been a call for mentoring or induction programmes for early career teachers (Haynes, 2014), there is still limited discussion about how schools can foster an environment that empowers early career teachers to strengthen and reshape their identities rather than narrowly identifying areas of weakness that requires development.

Given the realisation that to address early career teacher attrition there needs to be a focus on support, Burke et al. (2015) quantified early career teachers' preferences with regard to the various approaches to support. Interestingly, they noted that shared resources and collaboration with colleagues were valued by early career teachers and often lacking in schools where teachers were more likely to leave. Collaboration and sharing of resources not only allow early career teachers to learn from more experienced colleagues, but allows them to work co-operatively within a group and take an active role in working with others. This legitimises them as members of the group, while also providing both opportunities to learn and develop and providing some aspects of leadership as well.

Negotiating the school context or field

While teacher attrition has been linked to a lack of support, a number of scholars have argued that this problem is linked to the structure of school contexts (Marc-Bujosa, McNeill, & Friedman, 2019; Pietsch & Williamson, 2010; Stacey, 2019). While many scholars have attempted to understand early career teachers' experiences based on the pre-defined stages of one's teaching career, many scholars find this unhelpful due to the variability of school contexts (Day, 2017; Pietsch & Williamson, 2010; Stacey, 2019). As early career teachers attempt to negotiate their context in order to become legitimate members of their new teaching community, Stacey (2019) suggests that some teachers are better suited to some school contexts than others. She contends that there are 'issues of system-wide stratification' within the schooling system (Stacey, 2019, p. 405). This stratification influences how well a teacher, particularly an early career teacher, fits within their school context. She illustrates this, as she draws from interview data from an early career teacher in her study:

> Tim … described having come to develop a view that some teachers were better suited to particular schools than others, and that if a teacher was 'wrong' for a particular school, they simply would not 'survive'. Arguably, Tim's middle-class habitus may have been in almost 'radical disjunction' (Bourdieu & Wacquant, 1992, p. 130) with the schooling field in which he worked, requiring him to resource a range of extensive and to some degree unfamiliar socio-cultural, creative and relational demands.
>
> *(Stacey, 2019, p. 413)*

This excerpt highlights how a teacher's habitus can be in disjunction with the schooling field. This suggests that if a teacher's ways of thinking and doing, or habitus, are not aligned or appreciated by the school, this may result in a teacher feeling disempowered. Similarly, if the teacher is constantly questioning their own identity and struggling to find harmony within, then that leads to further disenchantment too. Stacey (2019) argues that it is not necessarily about the teacher being wrong for the school, but reveals what is wrong with the system. Therefore, school context plays a large role and has a powerful influence in developing and empowering early career teachers.

The teacher's stories: a three-dimensional analysis

While this study focuses primarily on the experiences of one participant, Waruni, in her suburban school, the perspectives of other participants in this study are included, as the group attempted to collectively unpack or make sense of Waruni's experience in light of their own. For the purposes of this chapter, Waruni's school is typified as a 'hierarchical' school. The school has been defined in this way due to Waruni's perspectives on how early career teachers are positioned within this school, denoting a hierarchy or stratification where novice teachers do not lead, but follow. Early career teachers are positioned as receivers of the school leadership's vision, policies and administration rather than active agents (Ball, Maguire, Braun & Hoskins, 2011), who have the school's support to lead others in ESE initiatives. Our three-dimensional narrative inquiry analysis investigates the temporal, societal and place-based dimensions of these teachers' stories to illustrate how early career teachers are positioned in schools as they attempt to lead ESE initiatives. First, the three-dimensional analysis reveals how Waruni grapples with her positioning within the school and how she comes to terms with how to create meaningful change that is supported by her school over time. Second, the analysis illustrates the politics that exist within schools, particularly when it comes to teachers leading other teachers. Finally, communities of practice, where early career teachers can be legitimised through discussions with like-minded people, create valuable spaces or places for early career teachers to feel supported as they navigate their early career teaching experiences.

Small steps over time create change: *'Baby steps... I wish I had known that before'*

By interrogating Waruni's stories through the temporal dimension, the findings suggest that passion must be packaged and handled to suit the context. In order to play the game and navigate the hierarchies of her school context, Waruni had to repackage her boundless passion for ESE into small, slow and intentional changes over time.

From the first workshop, Waruni's passion for ESE was evident, as she had a visible desire to ensure that her convictions about sustainability and the environment created meaningful change in herself and others. It was clear that Waruni's habitus was shaped by what she has learnt about sustainability and the environment at university, so much so that it not only impacted her ways of thinking but also her actions. After being inspired in one of her undergraduate units (a unit led by one of the researchers), Waruni chose to become a vegetarian. She explained this during the first workshop, as she described how a suggested book chapter changed her life:

> I just read 20 pages and that was enough for me to [think] oh my goodness how selfish am I to be doing the things I was doing and not really thinking about how my actions affect the global community. So, it was more internal, made me question the things I am doing, and really made me investigate sustainability first. That was my starting point.
>
> *(Waruni, workshop one)*

> Quite radically actually…I remember you came and you told me that you became vegetarian.
>
> *(Researcher, workshop one)*

> I thought, there are small things I can do in my life to have a positive impact towards this cause.
>
> *(Waruni, workshop one)*

Waruni, therefore, took to heart what she had learnt about sustainability in a way that encompassed her ways of being. However, while Waruni saw that change started with herself, she also acknowledged that her passion for ESE could have an impact on the learners in her classroom:

> It helps having a passionate teacher, that's going to show them [the children], that it is important.
>
> *(Waruni, workshop one)*

Despite her passion, however, Waruni found that she had to navigate this passion in a way that best suited her school's context and/or the existence of perceived hierarchies. She had tried to lead several ESE initiatives in her school, but there was always pushback. Feeling frustrated and disempowered, one of the community partners, provided Waruni with some advice:

> You've got to be aware that some teachers have never worked outside of a school or an academic environment and so they have a very insular view of those sorts of things, and things can get incredibly petty. My advice would be that you always take baby steps and you find out where the hierarchies are and you find out what you need to do to get approval before you jump. And I know that's after the horse has bolted. And learn that you have to play the game a little bit because ultimately what you want is the objective; you want to achieve your objective. How you go about it doesn't really matter, as long as you get what you want in the end and principals are always a good place to start.
>
> *(Community partner, workshop two)*

> Baby steps… I wish I had known that before.
>
> *(Waruni, workshop two)*

This advice resonated strongly with Waruni as she had tried to lead initiatives as an early career teacher (as discussed in more depth later), but did not realise that she needed to have support, backing and approval to validate what she thought would create meaningful change within the school. The idea of taking baby steps was a key message for Waruni, a way to test the waters—to navigate the school hierarchies in a way that worked for her rather than against her. The community partner's advice about 'you have to play the game' aligns with Bourdieu's concept of field. There are players in this game (Bourdieu, 1977) who hold a great deal of power, or capital, and they often play an important role in creating a collective habitus in school. Unfortunately, early career teachers within the schools must seek out the support and backing from the powerful actors, such as the principal, and ensure that they are playing by the unwritten rules of the game. For early career teachers, it can be hard to identify what the rules are. The community partner suggests that one must 'play the game a little bit.' This allows one to get a feel for the game, so you can understand not only the rules but power structures that you need to use in order to meet your objectives. As Dalal (2016) suggests, as he unpacks Bourdieu's work, the social contexts provide 'cues and impart logic to an individual' (p. 235) which must be heeded in order to be legitimised within the field. Therefore, Waruni needed to take the time to look for these norms, cues and power relations that shaped her school context.

The idea of playing the game for the sake of getting what you want seems ingenuine; however, when you are trying to play the game to create change, it matters to take slow, small and cautious, yet intentional, steps towards meeting one's goal. Waruni began to internalise these ideas in an attempt to persevere and not feel disempowered by her school context, as realised in her comments made in the following workshop a month later:

> It's just about starting small, not getting disheartened if something doesn't pan out and persevering. And having a plan in place I guess, before you go guns blazing. So small things affect big change.
>
> *(Waruni, workshop three)*

When someone is passionate about something and their convictions are strong, particularly if these convictions can fuel meaningful change, it can be hard to contract these powerful and boundless passions into small, procedural actions. Deep-set personal and environmental identities, when not matched with support to enhance professional identities for a harmonious confluence, can be frustrating for individuals. However, Waruni began to realise that big change starts small and that passions must be intentional and be navigated over time:

> I just found from the last session that starting slow, starting small and just creating a buzz about whatever initiative you wanted to put into place rather than all guns blazing. So, I found that really useful, that step by step

process, and basically applying marketing strategies to EfS. I thought that was really useful and I wish I had known that before, like the very, very first session, I think that would've been beneficial...

(Waruni, workshop three)

Waruni's application of marketing strategies highlights how ESE must be marketed to schools and teachers in a way that they can have personal buy-in and come to the realisation that they want and need this particular initiative. Many resist changes when they are told to do so without feeling as if they had a say in the change. Using this concept of marketing that was introduced by one of our community partners (see Chapter 6), Waruni began to realise that she could only peak others' interest by creating meaningful change in the things she did in her classroom ('creating a buzz') that could inspire other teachers to follow her lead rather than trying to lead teachers who were not willing to follow. Waruni had not realised the politics that can exist in schools and the precarious position that early career teachers can find themselves in when they want to lead but find no one will follow.

The politics of teaching

In understanding Waruni's stories as she describes her experiences as an early career teacher in her school, the societal dimension of these stories is exposed as Waruni begins to see the 'hierarchy' of school leadership and attempts to reconcile her desire to share her ideas, but also acknowledges that she is a beginning teacher. As discussed in the previous section, Waruni came to the realisation through the course of the community of practice workshops that she needed to promote ESE slowly and in small steps. Drawing upon marketing concepts, Waruni realised that she needed teacher and leadership buy-in and that this might be accomplished through strategic small steps that created a 'buzz' rather than a bang. Her realisations came after one of the community partners had presented about how to create buy-in (see Chapter 6) in the second workshop. This prompted Waruni to share her story:

> That actually did happen to me, I did send out a few emails to the staff of the school just as awareness about where they're placing their rubbish. One of the big things I noticed was coffee cups, people were putting them in the wrong bin and so I just came on air and said, 'We shouldn't be putting them in the recycling bin, unfortunately it has to go to landfill' and I sent some links just to educate the staff and let them know, just to give them more information because they're probably not going to seek it out themselves. That was not perceived in the most positive light. I guess that was my fault because I didn't know there was a hierarchy that I had to go through, I didn't realise I had to seek out permission before I did that. So, it wasn't me trying to say you have to do this, I just thought I'd share some information that I came across and it shut things down for me a little bit.
>
> *(Waruni, workshop two)*

While Waruni had attempted to simply 'share some information,' this was appreciated by her fellow colleagues. As discussed earlier, Waruni was not only new to the school and an early career teacher, but she had not taken the time to get a feel for 'the game.' She had not realised there was a 'hierarchy' that she had to go through in order to share her ideas. While Waruni's intention was to share ideas that she felt were important, in what she thought was a community of colleagues, this experience 'shut things down' for her. She took a step in trying to lead her colleagues, misunderstanding that she was not positioned to do so, and was left feeling disappointed and disempowered. Therefore, while it is argued that early career teachers have the capacity to lead others through evidence-based knowledge and approaches and their own experiences (Byrne et al., 2018; Hume, 2013), this example shows that this needs to be understood within the particular school fields. While some schools might welcome early career teachers' ideas and experiences, and see them as potential leaders, Waruni's story highlights that how leadership is viewed and recognised in schools exposes the positions of power that have come to exist. Some schools may see leadership from a hierarchical perspective, while others might see leadership as something that can be cultivated at all levels. Kraft and Papay (2014) argue that school contexts play an important role in developing and maintaining teachers, particularly as school leaders make the decision to actively establish opportunities for teachers to collaborate and/or share their ideas and experiences.

To better understand Waruni's story over the course of the community of practice workshops, it is helpful to explore the experiences and stories of the other participating teachers. The societal dimension of these stories is exposed as several of the participating teachers commented on how, from their perspectives, it appeared that teachers did not like being led by other teachers, particularly early careers teachers. This made it difficult for teachers, like Waruni, to share ideas and lead other teachers in thinking about and acting upon ESE concepts:

> Teachers leading teachers, some can be funny about this whole hierarchy thing and it really upsets me that you've got to go through this and you know. There's politics in any work environment…
>
> *(Daniel, workshop two)*

> It's probably because they've had to jump through all these hoops to get to where they are so [they]want us to do the same… I suppose there are principals out there who might be big on history or they're big on health and it's like, "No we don't worry about that because we're big on this" rather than letting the teachers teach what they're strong in.
>
> *(Shaun, workshop two)*

Daniel specifically chooses the word 'hierarchy' in trying to respond to Waruni's story. He acknowledges that there are politics in any work environment, yet as a colleague he understands that this is upsetting. However, positioned as beginning teachers, they accept this as being part of the game or social context, acknowledging that they are positioned on the lower rungs of the hierarchy and must concede to being tasked with getting through it. Shaun suggests that this is a rite of passage, the need to start at the bottom of the hierarchy so one appreciates the leadership opportunities later. This, therefore, suggests that it is not about developing teachers and empowering them to be leaders at the early stages of their career, but encouraging them to bide their time and play the game until it is their turn. Unfortunately, this sentiment was echoed by a community leader who joined the ThinkTank for workshop six:

> *They* [early career teachers] *are too busy surviving the rigours of being in the class-room and all the other demands placed on them…new teachers cannot do anything…*
> *(Community leader, workshop six, ThinkTank)*

In other words, early career teachers should not focus on anything other than settling into the profession. While the community member most likely intended to protect early career teachers knowing that these first years are difficult, her arguments support concerns about falling into place within the school hierarchy and biding one's time. It also suggests that early career teachers are incapable of implementing ESE in their classrooms, while also managing the daily "rig-ours" of teaching. The community leader's comments demonstrate how, in an attempt to protect early career teachers from the demands of their first few years of teaching, their sense of self, or their confluence of identity, should be denied. We, the authors, argue that we must move past this type of rationalisation. Instead of arguing that early career teachers should not 'do anything' beyond what is required—which suggests that the problem is the teacher, not the structure of schooling which positions them in this way (Stacey, 2019)—we need to empower teachers to find that confluence and consider how we can change the system.

Shaun also points out that schools have their own institutional identities, or habitus, which can sometimes limit teachers' teaching to their strengths and drawing from their own experiences. He reveals that at times teachers' strengths may not be valued, particularly if they do not align with the normalised priorities, practices and values of the particular school context. This may not be a problem with the teacher, but a problem with the system (Stacey, 2019). Additionally, it does not provide a context of mutual trust and respect (Bryk et al., 2010) where early career teachers can develop their identities. Waruni's stories highlight that early career teachers must navigate their teaching environment, learn the rules—or the politics—of the game and try to find a space where they can be true to themselves, yet not disruptive of the normative practices that shape the school context. From their experiences, these early career teachers either learn to retreat and bide their time or they find a way to negotiate the normative practices in their school in a way that allows them to thrive.

Creating meaningful communities of practice

In exploring the place-based dimensions of Waruni's stories, it is revealed how a sense of shared space as a community of practice is an important part of early career teachers' development. The Community of Practice workshops revealed how discouraging it can be for early career teachers when they attempt to share their convictions with colleagues and not have these ideas or practices be legitimised. As Waruni explained, 'it shut things down for me.' Despite these disappointing and trying times, however, our study highlights how important it is to have a sense of place or a sense of community as one tries to understand and unpack their teaching experiences. Given that many scholars have explored the *resilience* of early career teachers (Johnson et al., 2014; Peters & Pearce, 2012), this, in itself, suggests that these early years of teaching are difficult for many teachers. While the intention of this project was to explore how teachers embed ESE in their classroom practices, its greatest impact was realised in the community it created. While the intention was to create a community of practice, the data continued to reveal how important a like-minded community is. In particular, after Waruni had shared her story in workshop two and other teachers had responded, she commented:

> But it is great having these conversations because I always leave feeling energised after having a chat with like-minded people. So that's great, especially if you've tried something in your school and then it hasn't worked out, it's like a dopamine boost, makes you feel good coming here knowing that there's people out there that think the way you do and are passionate about this.
>
> *(Waruni, suburban school, workshop two)*

In many ways, the workshops were an opportunity for these early career teachers to freely share and feel supported in doing so. When Daniel reflected at the end of the project, he realised that implementing ESE in his school, even while focusing on small changes, was much harder than expected:

> To be honest, when trying to make small changes [in school practices], I thought more people would be like-minded around sustainability - which on the surface they seemed to be. I realised that it had to be easy though and require little to no extra workload. In an already crowded curriculum with all areas fighting for time, if it was perceived hard or not a directive from the boss, then it would struggle to get traction.
>
> *(Daniel, reflective writing)*

While Daniel found a way to link ESE to school priorities, he realised that this was due to 'having the right conversation at the right time.' Not only did this 'like-minded' community become a place for teachers like Waruni to be recharged after trying something that did not work out, but it also became a space for sharing ideas and experiences that could be tried in other teaching contexts. Many of the participants were eager to implement ideas, but needed each other's support and ideas to spur them on.

However, it is important to note that this community of practice began before our research project began. These teachers' ESE journeys began much earlier, and either began or were confirmed when they engaged with and voiced their convictions in their teacher education sustainability unit. As discussed earlier in the chapter, Waruni's habitus and a significant part of her personal as well as professional identity—her ways of thinking about ESE and her decision to become a vegetarian—was the result of a university sustainability unit. This unit forged an important relationship, connecting the researchers with the participating teachers, but also an important relationship to the concepts of ESE. This project was created based on like-mindedness and matching environmental identities. It was created for and by those who were passionate about ESE, had strong environmental identities and were tasked with making sense of how their convictions about ESE translated into meaningful teaching practices through their professional identities.

Looking to the future: hopeful applications

While Waruni's stories highlight the difficult position that early career teachers can often find themselves in when they attempt to share their ideas and practices, they are also reminded of their positioning within the hierarchy. However, there are still hopeful applications that we can learn from Waruni's stories. The first is that we can find nourishing places to share and connect, even if these exist beyond one's school context. Second, these stories highlight the role of resilience and the drive to learn from our experiences and to keep trying.

In Waruni's desire to connect and share with colleagues about her passion around ESE, she realised that there were ways to go about this so that she could be true to her convictions, but also understand and negotiate the structure and hierarchies established in her school. Waruni learnt that she did need to take the time to learn the rules of the game—particularly whose backing and support she needed so that her ideas and practices were sanctioned. In negotiating these unwritten norms and starting small, Waruni might be able to share her ideas and practices. However, in the meantime, she needed to be surrounded by like-minded people who could not only understand her positioning as an early career teacher but also comprehend and share her convictions. This community of practice was important for developing and confirming Waruni's confluence of identities, where she found encouragement and opportunities towards a greater sense of harmony.

These stories highlight that early career teaching can be difficult, particularly because it takes time to adjust to this new social context. While some school leaders still need to realise the opportunities that they have to retain teachers by respecting their strengths and allowing them opportunities to share and lead without fear of failing, in the meantime, it is important for like-minded educators to support one another. The sharing of stories that occurred in these community of practice workshops allows early careers teachers to hold onto hope, knowing that they are not alone and that it is a learning journey that requires perseverance through taking small and strategic steps towards their goal of implementing meaningful ESE practices. In addition, given that it is widely acknowledged that the first years of teaching are difficult, it is common that early career teachers are discouraged from leading and sharing their ideas in meaningful ways but just to focus on managing the tasks at hand. While we, the authors, agree that small and intentional steps are important, we also acknowledge that the ways in which early career teachers are positioned is a systemic issue that lies within Australian schooling. However, we remain extremely hopeful, as this study allowed us to observe a group of early career teachers who are ready and willing to lead and create change and who are actively navigating their school contexts to do so.

References

Arnup, J., & Bowles, T. (2016). Should I stay or should I go? Resilience as a protective factor for teachers' intention to leave the teaching profession. *Australian Journal of Education, 60*(3), 229–244. doi:10.1177/0004944116667620

Australian Council for Education Research. (2020). *Literacy and numeracy test for initial teacher education students.* Retrieved from https://teacheredtest.acer.edu.au

Australian Government Department of Education and Training. (2016). *Quality schools, quality outcomes.* Canberra: Australian Government.

Ball, S., Maguire, M., Braun, A., & Hoskins, K. (2011). Policy actors: Doing policy work in schools. *Discourse, 32*(4), 625–639. doi:10.1080/01596306.2011.601565

Bourdieu, P. (1977). *Outline of a theory of practice.* Cambridge: Cambridge University Press.

Bourdieu, P., & Wacquant, L. (1992). *An invitation to reflexive sociology.* Chicago: University of Chicago Press.

Boyd, D. J., Grossman, P. L., Lankford, H., Loeb, S., & Wyckoff, J. (2009). Teacher preparation and student achievement. *Educational Evaluation and Policy Analysis, 31*(4), 416–440. doi:10.3102/0162373709353129

Bryk, A. S., Sebring, P. B., Allensworth, E., Luppescu, S., & Easton, J. Q. (2010). *Organizing schools for improvement: Lessons from Chicago.* Chicago, IL: University of Chicago Press.

Burke, P., Aubusson, P., Schuck, S., Buchanan, J., & Prescott, A. (2015). How do early career teachers value different types of support? *Teaching and Teacher Education, 47,* 241–253. doi:10.1016/j.tate.2015.01.005

Byrne, J., Rietdijk, W., & Pickett, K. (2018). Teachers as health promoters. *Teaching and Teacher Education, 69,* 289–299. doi:10.1016/j.tate.2017.10.020.

Carey, A. (2020, February 18). 'Like being thrown off of a bus': Mentoring program to ease load on teachers. *The Age.* Retrieved from https://www.theage.com.au/national/victoria/like-being-thrown-off-a-bus-mentor-program-to-ease-load-on-teachers-20200218-p541zo.html

Clandinin, D. J., & Connelly, J. (2000). *Narrative inquiry: Experience and story in qualitative research.* San Francisco, CA: Jossey-Bass Inc.

Dalal, J. (2016). Pierre Bourdieu: The sociologist of education. *Contemporary Education Dialogue, 13*(2), 231–250.

Day, C. (2017). *Teachers' worlds and work: Understanding complexity, building quality.* London, UK: Routledge.

Dover, A. (2018). Your compliance will not protect you: Agency and accountability in urban teacher preparation. *Urban Education.* doi:10.1177%2F0042085918795020.

Haynes, M. (2014). *On the path to equity: Improving the effectiveness of beginning teachers.* Washington, DC: Alliance for Excellent Education.

Hume, A. (2013). Early-career teachers as future agents of change in New Zealand primary science. *Journal of Educational Leadership, Policy and Practice, 28*(2), 3–14.

Johnson, B., Down, B., Le Cornu, R., Peters, J., Sullivan, A., Pearce, J., & Hunter, J. (2014). Promoting early career teacher resilience: a framework for understanding and acting. *Teachers and Teaching, 20*(5), 530–546. doi:10.1080/13540602.2014.937957

Kraft, M. A., & Papay, J. P. (2014). Can professional environments in schools promote teacher development? *Educational Evaluation and Policy Analysis, 36*(4), 476–500. doi:10.3102/0162373713519496

Marc-Bujosa, L., McNeill, K., & Friedman, A. (2019). Becoming an urban science teacher: How beginning teachers negotiate contradictory school contexts. *Journal of Research in Science Teaching, 57*(1), 3–32. doi:10.1002/tea.21583

Peters, J., & Pearce, J. (2012). Relationships and early career teacher resilience: a role for school principals. *Teachers and Teaching, 18*(2), 249–262. doi:10.1080/13540602.2012.632266

Pietsch, M., & Williamson, J. (2010). 'Getting the pieces together': Negotiating the transition from pre-service to in-service teacher. *Asia-Pacific Journal of Teacher Education, 38*(4), 331–344. doi:101080/1359866X.2010.515942

Stacey, M. (2019) 'If you're wrong for the place you just don't survive': Examining the work of early career teachers in context. *Teachers and Teaching, 25*(4), 404–417. doi:10.1080/13540602.2019.1621828

Weldon, P. (2018). Early career teacher attrition in Australia. *Australian Journal of Education, 62*(1), 61–78. doi:10.1177/0004944117752478

11

ESE IN A RURAL SCHOOL

'We became the grade who does things'

Box 11.1 Climate change making drought worse

The National Farmers' Federation has declared that climate change is making the drought worse in Australia and says tiptoeing around the subject does not do regional communities any good.

(Chan, 2018, para. 1 & 2, The Guardian, 6 October 2018)

Box 11.2 Extract from Eddie in workshop three

I've got kids that are in reading recovery type groups but I've timed my timetable so that they're taken out at the time we're doing reading so that they're not missing the other stuff, for that reason I don't want them to miss the other important stuff, like sustainability…

(Eddie, workshop three)

Introduction

While only one of the participants in the study was teaching in a small rural school, this early career teacher's (Eddie's) stories were important because they provided a case study on how reframing your perception of the school and community context can have a surprising outcome. As the project evolved, a mismatch became evident between how Eddie had assumed her school community would respond to the introduction of ESE in the classroom and what actually happened. While dominant discourses fuelled by political agendas and the media (Ulmer, 2016) may be responsible for the way farming communities are perceived; other media outlets, such as depicted in Chan's (2018) article in *The Guardian* abstract above (Box 11.1), go beyond the 'single story' (Adichie, 2017) of the conservative Australian cattle farmer. Similarly, Eddie's stories illustrated the contrast between the rhetoric and the reality of rural schools and their communities. This was seen in Eddie's surprise when her learners, their farming families, the Principal and her fellow teachers responded in positive, engaged ways and eventually followed the lead her class provided. From a group of learners which Eddie had assumed would not be interested, nor have any ideas about sustainability, they became the 'grade that does [sustainable] things' in a small school in rural Victoria.

This chapter begins with a review of the literature around sustainability in regional and rural educational settings, and the important role of place-based learning in meaningful curricula. Following this, the early career teacher's stories from the community of practice conversations are presented, positioned within the tenets of the three-dimensional analysis. These stories and their analysis includes the temporal dimension with shifts in attitudes about ESE over time; the societal dimension and the influence of media on teachers and learners; and finally, the place-based dimension, with the question raised about rural schools being perceived as overly traditional rather than a space for opportunities. The final section of this chapter provides hopeful applications of ESE ideas and suggestions that may be adapted into rural classrooms in the future.

Regional education: inadequate or poised for opportunity?

Many argue that education within rural and/or regional areas is often unjustly framed as being deficient or disadvantaged (Barnes, 2020; Green, 2015; Masinire Maringe & Nkambule, 2014; Reid, Green, Hastings, Cooper & White, 2010), ignoring the strengths and opportunities that regionality provides. For example, Barnes (2020) argues that given the strong community networks which exist within regional and rural areas in Australia, there are opportunities for *local* approaches to regional educational problems and issues rather than attempting to adapt urban approaches to manage local needs. In the case of Africa, Masinire et al. (2014) contend that there is a binary between urban and rural education, defined by difference in regard to both opportunity and adequacy, with rural areas positioned as 'objects of exploration awaiting philanthropic and exotic interventions' (p. 148). They argue that when rural areas are viewed in this way, it disregards the strengths and opportunities that exist within the community's local knowledge, histories, languages and cultures. Furthermore, this ignores the individual differences among regional and rural communities within a particular area (e.g. country, state, region) and how they are uniquely defined by their specific practices and spaces.

The notion of regional and rural areas being defined by their strengths and opportunities rather than by the points of difference when compared to urban areas is important, particularly in light of the value of place-based education. As discussed in Chapter 6 on the community partners, place-based learning embraces the role of *place*, particularly when viewing place as one's local community and environment, as a valuable and authentic space for learning. As mentioned earlier, connecting to nature within one's local community can motivate environmental behaviours as it allows young people to build their environmental identities, extend their connection and love for spaces to environmental problems and concerns (Pettifer, 2019). Applying concepts of biodiversity and agriculture that were linked to the local farming community allowed local school children in China to connect their learning with the authentic environmental concerns which related to their nearby reserve (Chan, Mathews & Li, 2018). Therefore, these concepts took on new meanings when realised within the school children's community context. Additionally, the community came together to understand and identify environmental problems and concerns.

Rural-regional sustainability

There is a strong connection between rural and regional areas and sustainability, particularly given that metropolitan areas in many countries rely on the rural and regional areas to sustain and provide food, energy and water to the rest of the country (Halsey, 2011). In discussing this connection between rural-regional areas and sustainability, Green (2015) proposes the term *rural-regional sustainability*. This term acknowledges that (1) regional and rural are not synonymous but distinct; and (2) there are strong connections between rural-regional areas and sustainability. He argues that rural-regional Australia, in particular, has a unique mix of demography and geography (Green & Reid, 2004), which has unfortunately seen a decline in population and natural resources in light of increasing environmental stress and concern about climate change. Another interesting characteristic of rural-regional sustainability that is emphasised not only in countries such as Australia but also in less developed countries in Africa (see Masinire et al., 2014) is the focus on *regeneration*. The term regeneration has been used to acknowledge the suppression, disorder and painful histories that characterise many rural and regional areas (Green, 2015; Main, 2005; Masinire et al., 2014; Reid et al., 2010). This term emphasises that in these areas, it is not only about sustaining but also acknowledging that these areas have been oppressed and require a focus on (re)development:

> Africans were subjected to an inferior education, whether in what was then termed townships or in underdeveloped homelands. Thus, rural education resurrects the past injustices of the homeland policy which was particularly invidious in relation to resource allocation and governance structures...
>
> *(Masinire et al. 2014, p. 149)*

Writing in the specific context of Australian environmental and agriculture history, with its long (mis)engagements with Indigenous experience, Main's (2005) argument has resonance and relevance for other (postcolonial, post-settler) countries as well (Green, 2015, p. 37). Therefore, for rural-regional areas, a focus on sustainability requires a focus on development.

The relationship between rural-regional *development* and sustainability is a matter of concern in both developing countries and the wider international community and requires a distinctive place within education curricula (Halsey, 2011; Masinire et al., 2014; van Crowder et al., 1998). Whether food security, energy and/or water, rural and regional areas are often key to sourcing, producing and/or managing these resources for urban consumption. However, rural-regional development is often not prioritised, particularly in light of the impact of climate change (Halsey, 2011). The livelihood of many rural and regional areas is dependent on their unique spaces, and in the case of Australia, what these areas have to offer sustains the rest of the country. Green (2015) suggests that there is 'widespread anxiety over the fate and fortunes of rural Australia, in a global context of climate change and ecological challenge' (p. 36).

Despite this anxiety, Halsey (2011) highlights the vital role education plays in the relationship between sustainability and rural development, claiming that education is important for 'nurturing and releasing capacities in imagination to construct, energise and drive what is necessary to transition our nation to a "sustainable state"' (p. 64).

Education plays a key role in the regeneration of rural-regional communities and is vital for the social, economic and environmental sustainability of these areas (Reid et al., 2010). Halsey (2011) highlights the need to release capacities to transition to a sustainable state. We, the authors, argue that this can only be accomplished if ESE holds a position of knowledge power within schools. However, there are a number of challenges that regional areas face in ensuring that the learners receive a strong induction to ESE, particularly given the challenges in recruiting and retaining teaching staff (Cuervo & Acquaro, 2018; Halsey, 2018: McKenzie, Weldon, Rowley, Murphy & McMillan, 2014; Reid et al., 2010). Given the contested role that ESE plays within the curriculum and the challenges that metropolitan teachers face in employing the multitude of teaching materials and resources currently available (Barnes, Moore & Almeida, 2019), the added constraints of limited professional development opportunities and access to resources and funding that many rural and regional schools face (Barnes et al., 2019) make the implementation of ESE in these schools increasingly problematic. However, strategic education strategies, which focus on rural and regional development, have the ability to democratise the curriculum rather than restrict it (Masinire et al., 2014). Making clear ESE links within the curriculum allows learners to identify and understand the local problems and solutions that connect with their sense of place. By embracing place-based learning (Pettifer, 2019), schools can make ESE meaningful and relevant to learners, while also 'nurturing and releasing [their] capacities' to redevelop and regenerate rural and regional areas (Halsey, 2011, p. 64).

The teacher's stories: a three-dimensional analysis

The three-dimensional narrative inquiry analysis process, used to identify findings in this study, highlighted the temporal dimension as the tension between what Eddie had assumed would occur in her classroom if ESE was introduced and the changes that happened over time. In the societal dimension, the influence of the media on learners' understanding of sustainability issues was made visible; while the notion of empowering learners through ESE became increasingly evident. The place-based dimension was significant for Eddie, as a new teacher in a small rural school with strong environmental identities, and her learners who responded so 'excitedly' when given the opportunity to regenerate their own local environment.

'A taboo thing': the perception of older teachers not valuing ESE

A key narrative theme identified in the analysis of the temporal dimension in Eddie's stories was her assumption that the *older teachers* she worked with in the rural school did not value or want to engage with ESE. Eddie considered this was because they deemed sustainability *a bit of a taboo thing* and therefore, not to be included in the classroom, as depicted here in this story extract below from workshop one:

> I think it's like a passion thing as well though, I mean if you're the only teacher there that does care about [sustainability] then…do you have to educate others? A lot of the teachers I work with are much older, and sustainability is a relatively new concept in a lot of ways you know… so they wouldn't have had any at all really, it wouldn't have been part of their education [as teachers], and it's very much viewed as a bit of a taboo thing more so in that age group I'm working with, and I wonder whether it's kind of a upward battle whether it's in the policy or not. You've also got the battle of getting people to actually believe that it is necessary and worthwhile.
>
> *(Eddie, workshop one)*

It is noteworthy that Eddie made such assumptions about the *older teachers* she worked with, seemingly based on her perception that no ESE was visible in the school and therefore sustainability was not a topic listed to be discussed at teacher meetings, with school leadership nor within the curriculum. Eddie's stories show she felt somewhat overwhelmed that she was the 'only teacher there that does care about sustainability,' and wondered further if it was therefore her responsibility to 'educate others.' This highlights the disjunction between her habitus and the collective habitus of the school. This dissonance forced her to question as to whether it was her role and whether she was suitably positioned to shape other teachers' ways of thinking and/or responding to sustainability and consequently shape the social field. Eddie also felt concerned it might appear she was imposing her environmental identities onto others. The stories and further analysis in the temporal dimension that follows indicate how far Eddie was able to *educate others* along her own pathway into the enactment of ESE with her learners.

A shift in thinking from *'I don't know what to do'* to *'We can do it'*

In examining Eddie's stories over the duration of the project, it is interesting to note how dramatically these stories changed in their tone, their content and their degree of enthusiasm. Within the analysis of the first temporal dimension of Eddie's stories, these shifts were clearly illustrated over time. For example, in the first introductory workshop, Eddie openly revealed her university experience in the Sustainability Unit as being '... probably the first time I ever thought about sustainability at all actually. It just wasn't something that occurred to me as a thing.' Eddie's revelation continued:

> It just opens up your thinking, I think. And that's when you realise that you do need to change something and do something as well....my husband thinks I'm crazy since I'm at the beach now with my hands filled with bits of plastic. But you don't really think of it until you sort of actually see it....
>
> *(Eddie, workshop one)*

However, despite Eddie's declaration on the need to 'do something' at the beginning of the project, she was the first participant to disclose that she was not sure what to do to embed ESE into her classroom, saying, 'I'm not sure if I'm equipped still to know where to start.' However, by the end of the project, Eddie and her learners became the grade 'who do [sustainable] things... and the other kids in other classes would start looking up to them' (Eddie, workshop three). This was evident in workshop three when Eddie announced that not only were the learners in her class fully engaged in the concept and practices of ESE, they were eager to pass on important information to others in the school, even without Eddie's assistance, as explained here:

> ...and the buzz they got from when Bunnings actually delivered the bins, they thought it was brilliant and they were helping them bring them in from the car. So we've had these bins sitting in my room for the last week and a half because we need to prepare something so we can let everyone know what we're doing. In about an hour's time my kids will be presenting at assembly with a video we've made. So I've prepped them this morning before I left [to come to our community of practice meeting] and they're presenting what we're doing. Then on Monday we'll be delivering all the bins to each of the classrooms with the posters the kids have made to show what each one's for. So that's been really good.
>
> *(Eddie, workshop three)*

Demonstrated here in this story is the palpable sense of agency, problem-solving skills and confidence exhibited by these learners in front of their peers, other adults and teachers. In many ways, Eddie's own personal shift in thinking over time was manifested in her learners' experiences of ESE, from initially not knowing what to do to becoming the grade that everyone else literally followed. Ironically, in terms of the project, Eddie was the only participant who actually *did something* alongside her learners as we had originally envisaged for all the participants. Eddie's story emphasises that 'power is exercised rather than possessed' (Foucault, 1977, p. 26). While positioned as being the only teacher in the school who was passionate about ESE and her habitus seemingly in conflict with the collective or institutional habitus of the school, Eddie's decision to *do something* resulted in meaningful changes within her classroom and school. As Foucault contends, 'Furthermore, this power is not exercised simply as an obligation or a prohibition on those who "do not have it"; it invests them, is transmitted by them and through them…' (Foucault, 1977, pp. 26–27). In Eddie's story, she, an early career teacher, exercised power that was transmitted through them to empower her students.

The influence of media on learners' engagement with ESE

Through the societal dimension of the analysis, the influence of television documentaries and YouTube clips became increasingly evident. This was seen, for example, at the beginning of the project when Eddie told stories about her experience at university in the Sustainability Unit watching videos, such as the Great Pacific Garbage Patch (cf: https://www.youtube.com/watch?v=1qT-rOXB6NI), which triggered her emerging understanding about the need to 'change something and do something' in relation to sustainability, as seen here:

> And in there we watched some videos about the pollution in the ocean and some of those pictures were pretty awful, the animals that have been washed up on the beach had been cut open with stomachs filled with bits of plastic. And I think that was the same when we went into the third-year unit, same sort of thing, we were exposed to those ideas….
>
> *(Eddie, workshop one)*

Interestingly, it was these same YouTube clips and other media productions that Eddie decided to show her Grade 5 learners at the start of the research project, thinking it may provoke a similarly powerful response to her own experience, as seen in this story:

> It was something I don't think had been really looked at [in this school] before, but my Grade 5s are quite passionate about it now I think because I've shown them things along the way. I've shown them things like the Great

Pacific Garbage Dump and those sorts of videos that are aimed at shocking a little bit but just to educate....

(Eddie, workshop two)

The following story that Eddie told late into the community of practice conversations in workshop two illustrated how the ABC documentary *War on Waste* (Reucassal, 2017) sparked creative thinking and problem-solving amongst Eddie's learners. Starting her narration tentatively, Eddie's story gradually unfolded with the introduction into her classroom of an hour of *sustainability*, as she explained here:

So... I've just started by putting in an hour a week that I've just called sustainability, mostly because then at least I'm doing something because I'm still finding it quite hard to integrate. But I've got that hour a week at least.

(Eddie, workshop two)

From this hesitant start, Eddie continued on in her storytelling with increasing enthusiasm and, building in confidence, she happily reported:

So, the first day back we emptied our rubbish bin onto a black bag on the floor and sorted it like War on Waste into the four, and the kids couldn't believe that there was literally three things that needed to go to landfill out of the whole bin, most of it was compost. I've got a class of 30, so a nice big class, and half of them wanted to write to the Principal - I told them they had to write a letter that was my direction – but half of them wanted to write to the Principal to request that she initiate a Nude Food Day because we don't have any. So, I said to them, "Our biggest problem is we don't have enough bins" as in there's only one big bin in each room, everything goes in there, "We need more of a system" and they said, "We need four bins" and I said, "How are we going to get them?" and they said, "I wonder if anybody would donate them" and I said, "I think we would get donations. Let's write to Bunnings....

(Eddie, workshop two)

At this point in Eddie's storytelling, the rest of the project participants were excitedly supporting her progress, encouraging Eddie to explain more of how she had engaged her learners in ESE. With a flourish, Eddie announced to the group:

So we've just finished writing them, I've got them in my car ready to post over the weekend.

(Eddie, workshop two)

The role of the media can be seen to have a powerful influence on the way viewers interpret current debates (Thomas, 2011; Ulmer, 2016), such as debates around climate change. For Eddie, it appears her purposeful use of media was exactly the trigger she had predicted would prompt her learners to respond in positive sustainable ways. When asked in workshop three if anyone had any 'small steps' they would like to share with the group, Eddie no longer held back with her stories of success, and responded saying:

> Yeah, well, I've had quite a lot. I think last time I told you guys that we wrote letters to Bunnings. Well they delivered some bins to us the week before last, so we've got enough bins for every classroom as well as a school compost bin which is really good because the school just didn't have anything like it before.
>
> (Eddie, workshop three)

Also illustrated here in Eddie's stories over time was the development of inquiry-based learning embedded in ESE, with the co-construction of knowledge and problem-solving between the teacher and the learners. This was the underlying premise that unfolded between Eddie and her learners as they were increasingly able to create and/or envisage solutions to environmental problems in their local area.

'The grade who does things': empowering learners through ESE

In another theme identified in the societal dimension of analysis, the empowerment of learners became a dominant narrative pattern throughout many of Eddie's shifting stories and attitudes about her group of learners. At the start of the project, it was evident in Eddie's stories that she believed her Grade 5 learners *would not even know what sustainability means* and therefore, would not be interested in sustainability as something to discuss in the classroom. Regardless of the researchers' and community partners' conversations around the ESE work enacted with young children (see for example Chapter 8), Eddie was convinced during workshop one that her learners' world view on sustainability had already been *squashed down* by their farming families. In other words, she feared ESE concepts might be muted in this particular social field.

However, as previously revealed in this chapter, Eddie's earlier misconstrued perception of her learners did not materialise as they became agentive role models for sustainability for the other grades in their small rural school. This was evident in the following stories told by Eddie in workshop three, when she spoke of her learners' excitement and sense of *power* to *see something happen*:

> You should've seen my kids when the Bunnings woman walked into the room, they were just like bouncing out of their seats with excitement, it was crazy. I think it gave them some power. They didn't realise that if they wrote a letter requesting something that they might actually make a difference and see something happen. It took a long time, it took about six

weeks from when we wrote the letters to actually getting them. So again it's the small steps sort of thing…

That was one thing and it took a while to get to that point, and it's taken us another week and a half to be organised for actually presenting it to the school because we wanted to get it right, we didn't want to just dump them in the rooms and people don't know what to do with them. It's better to just get it right the first time and then hopefully everybody will be confident…

And then, Eddie proudly announced:

Yeah, so we were recording this morning – we're making a video application for a grant, we're asking for $10,000 from the Nude Food Company to fix up the school's wetlands area at the back, the kids' ideas include putting a dome over the top of it… so they had all sorts of great ideas… So we've made a video, we've got Grade 1, 2 and 5s together and I've got two other teachers on board with me for that…

(Eddie, workshop three)

Eddie's stories about the changes in her learners reinforce what other researchers have found with the empowerment of learners in and through ESE (Engdahl, 2015; Harley et al., 2018; Ruston, 2008; da Silva et al., 2020; UNESCO, 2018). For example, in Harley et al.'s study, school students across Europe engaged in critical thinking about the impacts and solutions around marine rubbish by participating in a video competition which highlighted environmental problems. As a consequence, the learners increased in their confidence, competency and readiness to act, while also 'fostering a sense of citizenship and ownership to give students an active voice in the problem and empower them to act' (Harley et al., 2018, p. 229).

Similarly, Ruston's (2008) earlier work found learners were more motivated to 'want' to learn about the local environment when they were given opportunities to be more actively involved in 'deep learning' about their world around them. Of particular relevance to Eddie's stories, da Silva et al. (2020) found what was important for ESE to succeed required school leaders and teachers to first acknowledge that young learners can act as 'change agents' in sustainability and, therefore, not to underestimate their learners' capacity as 'main actors' to make 'important decisions about the future…' (p. 6660); and second, the school leaders and teachers 'must be vigilant to make sure this conception lasts…' within the school community (p. 6661). With Eddie's increasing belief and confidence in her learners' capacity to understand and subsequently act on ESE issues, a wide range of learning opportunities around ESE became available for them to work on together constructively in a way she had not previously thought possible.

Perception of a backlash from farming families

In the analysis of the place-based dimension, Eddie's perception of her learners, their families and the school had been clearly impacted by the rural context where the school was located. Context is a critically important determinant in the design of any curriculum (Churchill et al., 2019); however, so too is ascertaining the learners' 'funds of knowledge' (Moll, Amanti, Neff & Gonzalez, 1992) and what they already know rather than making assumptions about their level of understanding and/or prior knowledge about particular content. In workshop one, Eddie was adamant that she needed to seriously consider the cattle farming background of her learners' families when she was organising the curriculum content, saying 'the parents will be a problem' when she was thinking about ESE for the research project. As a consequence, Eddie was convinced she would suffer a 'backlash' if she introduced sustainability as a concept, arguing:

> ...there would be a backlash from parents if the students came home talking about this...I need to take little tiny steps in my classroom, they [the learners] don't even know about sorting rubbish...'
>
> *(Eddie, workshop one)*

However, what Eddie had not initially taken into account was that her assumptions were based on traditional dominant discourses around rural farming communities rather than the reality of their contemporary attitudes to sustainability. Chan's (2018) article in *The Guardian* at the beginning of this chapter is testament to the more recent shifts in attitudes around cattle farming, with the National Farmers' Federation now acknowledging the need to address climate change to mitigate the long-term effects of drought. Further to this, the following extracts in Figure 11.2 below, from an Australian UNESCO programme that encourages primary school learners to work towards sustainability, is a stark reminder of the power of young learners in bringing about change in their own local communities, which states:

Box 11.3 Youth take the pulse of the planet and create change in their own communities

'What we do is to unashamedly focus on those people who can bring real change in their communities and the world and help them kindle that flame,' said Sue Lennox, co-founder and former CEO of the Global Rivers Environmental Education Network (OzGREEN)...

   The YLTW programme uses citizen science, education for sustainable development (ESD) and participatory leadership to inform, involve and connect young change makers within communities...

(UNESCO, 2 August 2019)

Sue Lennox, the co-founder of OzGREEN, spoke of the movement of people from 'inaction to a place where they can start imagining and creating a different world' through an understanding of what was required in 'their own region,' and her words are of further relevance to Eddie's and her learners' experience of ESE. Sue's comments mirrored exactly what Eddie and her learners had advocated for in the regeneration of the wetland area at the back of their regional school, as seen here with the learners' local knowledge coming to the foreground in their approach to regenerating the wetland area:

> And the kids said there's animals in there that might come into the wet-lands if it's a better environment. So, the kids want to build boxes for the birds and all that sort of stuff.
>
> *(Eddie, workshop three)*

Researchers have become increasingly aware of children's strong connection with place (cf: Hart, 1979; Moore, 1986; Rasmussen, 2004), and the 'special bond' within a 'sense of place' that children develop with the environment they grow up in (Measham, 2006, p. 3). More recently, researchers interested in ESE have found that learners and teachers engaged in 'authentic place-based learning' have been able to play a key role in the regeneration of local rural areas (Green, 2015; Pettifer, 2019; Reid et al., 2010). This phenomenon of learners' strong con-nection to their own local 'place' was evident in Eddie's stories of her learners' creative ideas around the regeneration of the wetland space, when she said:

> I had to bring [*the learners*] back to earth a little bit… and said, "Mainly we want a path that goes through it and we're looking at trying to put in an outdoor classroom." The biggest issue we'll have will be that we probably won't be able to use the space in summer because there are a lot of snakes because we're in the country, so there's a bit of an issue with that, but oth-erwise it'll be a really good space to use for the rest of the time. But it's very overgrown and hasn't been looked after so it's a fairly big job. So that's what we're in the progress of doing at the moment.
>
> *(Eddie, workshop three)*

What is notable in Eddie's stories here is not only around the level of excitement and commitment to their local 'place' exhibited by her learners, but Eddie's own emerging sense of purpose in the regeneration of the local area and her increased understanding of its significance to her learners.

Perception of a rural school—more traditional or possibilities and opportunities?

Eddie's earlier statements about the rural school and her position within it were indicative of her misunderstandings about the school's strong connection with local issues. As already discussed, Eddie had thought her school community would not be willing to engage in ESE, believing they held deeply entrenched, traditional beliefs that excluded ESE from the curriculum. However, encouraged by the research project to speak up to her Principal and find out what ESE opportunities may be possible, Eddie was able to move beyond her assumptions and begin to understand that one's habitus is shaped by, but also *shapes*, the social field. In other words, her habitus or ways of thinking and doing can have an impact on the institutional habitus of the school over time. As a consequence, Eddie discovered not only that the inclusion of ESE was possible, but how quickly it could happen, as seen in the Principal's immediate response below:

> I gave the letters to the Principal last night and I had a response the same night with a letter back for me to show the kids, she's also gone onto websites to search for what we can do next in terms of doing a Nude Food Day and getting some funding for it. So we're in quite a good position in terms of ESE projects now - I don't know how the rest of the staff will perceive it but me and the Principal are fine. The Grade 1/2 teachers are already doing it in their room as well so I know they'll be on board, I'm not sure about the others, we'll just have to see when it gets a bit further. I haven't really shared a lot with the whole staff... yet.
>
> *(Eddie, workshop two)*

Rather than what may be perceived as a more traditional approach to curriculum, what eventuated in Eddie's inquiry-based integrated project with her rural school classroom of 30 Grade 5s, was a similar experience to the findings in Chan et al.'s (2018) study. The learners in that study and Eddie's learners alike were each able to link the concepts of biodiversity and agriculture to their local farming communities through their work in a local reserve, as seen here:

> The problem in there at the moment is there's a lot of weeds and non-natives have overtaken a bit, so even if we consider it just an area for biodiversity – we've got an area about a kilometre away from school that Landcare have been working in and doing a lot of stuff so that's now available for us to use as well...
>
> *(Eddie, workshop three)*

As indicated earlier in this chapter, clearly linking the curriculum with 'local problems and solutions' connects learners and their teachers with their own sense of place and activates a deeper approach to experiential place-based learning (Pettifer, 2019).

Looking to the future—hopeful applications

Eddie's stories provide hope for early career teachers and teachers more broadly, as her experiences emphasise that changes can begin at the personal level (e.g. the decision to *do something*) and transform both the classroom, for example, through empowering the learners and shifting the timetable (as seen in the opening quote) and school practices, for example, with the principal's supportive actions, other teachers coming on board and in curricular change in the inclusion of ESE. It is a great example of the confluence of identities and the powerful changes that can be enacted when personal, environmental and professional identities work in tandem. While Eddie can be seen as initiating much of this ESE work within her classroom and then at the whole-school level, the Principal and other teachers were open to making significant changes in their curriculum. In workshop three, conversations about the creation of a 'buzz,' Eddie mentioned, '…yeah [the buzz] is really good, that's three out of seven of the teachers are involved now, so that's almost half of them [in my small school]…' With an increasing focus on the local environment with the local teachers, learners and families, Masinire et al.'s (2014) prediction that the curriculum can become democratised rather than restricted is illustrated here in Eddie's storied experiences. Additionally, Eddie's stories throughout the community of practice workshops highlight how meaningful change at the classroom and school level can occur, even when one is positioned as feeling marginalised by the ways in which they think about the world around them. Her experiences provide hope that not only can early career teachers lead change, but that powerful change can be transmitted and made meaningful for learners.

References

Adichie, C. N. (2017). TedTalk: *The danger of the single story*. Retrieved on 8th October 2020 from https://www.youtube.com/watch?reload=9&v=D9Ihs24Izeg

Barnes, M. (2020). Creating 'advantageous' spaces for migrant and refugee youth in regional areas: a local approach. *Discourse: Studies in the Cultural Politics of Education*. doi:10.1080/01596306.2019.1709415

Barnes, M., Moore, D., & Almeida, S. (2019). Sustainability in Australian schools: a cross-curriculum priority? *Prospects, 47*(4), 377–392. doi:10.1007/s11125-018-9437-x

Chan, G. (2018, October 6). All about the land drought shakes farming to its indigenous roots. *The Guardian*. Retrieved from https://www.theguardian.com/environment/2018/oct/06/all-about-the-land-drought-shakes-farming-to-its-indigenous-roots

Chan, Y-W., Mathews, N., & Li, F. (2018). Environmental education in nature reserve areas in southwestern China: What do we learn from Caohai? *Applied Environmental Education & Communication, 17*(2), 174–185. doi:10.1080/1533015X.2017.1388198

Churchill, R., Godinho, S., Johnson, N., Keddie, A., Letts, W., Lowe, K., Mackay, J., McGill, M., Moss, J, Nagel, M., & Shaw, K. (2019). *Teaching making a difference* (4th ed.). Milton, QLD: Wiley Publishing.

Cuervo, H., & Acquaro, D. (2018). Exploring metropolitan university pre-service teacher motivations and barriers to teaching in rural schools. *Asia-Pacific Journal of Teacher Education, 46*(4), 384–398. doi:10.1080/1359866X.2018.1438586

da Silva, A., Coelho, A., dos Santos, H., Neto, A., de Castro, A., & El-Aouar, W. (2020). Education principles and practices turned to sustainability in primary school. *Environment, Development and Sustainability, 22*, 6645–6670. doi:10.1007/s10668-019-00505-2

Engdahl, I. (2015). Early childhood education for sustainability: The OMEP world project. *International Journal of Early Childhood, 47*. doi:10.1007/s13158-015-0149-6.

Foucault, M. (1977). *Discipline and punish: The birth of the prison*. New York, NY: Pantheon Books.

Green, B. (2015). Australian education and rural-regional sustainability. *Australian and International Journal of Rural Education, 25*(3), 36–49.

Green, B., & Reid, J.A. (2004). Teacher education for rural-regional sustainability: Changing agendas, challenging futures, chasing chimeras? *Asia-Pacific Journal of Teacher Education, 32*(3), 256–273.

Great Pacific Garbage Patch. Retrieved from https://www.youtube.com/watch?v=1qT-rOXB6NI

Harley, B., Pahl, S., Holland, M., Alampei, I., Veiga, J., & Thompson, R. (2018). Turning the tide on trash: Empowering European educators and school students to tackle marine litter. *Marine Policy, 96*, 227–234.

Hart, R. (1979). *Children's special places: Exploring the role of forts, dens and bush houses in middle childhood*. Detroit, MI: Wayne State University Press.

Halsey, J. (2018). *Independent review into regional, rural, and remote education*. Retrieved from https://www.education.gov.au/independent-review-regional-rural-and-remote-education

Halsey, R. (2011). Farm fair voices, space, history, the middle ground and 'the future' of rural communities. *Education in Rural Australia, 21*(1), 39–66.

Lennox, S. (2019). Youth take the pulse of the planet and create change in their own communities. CEO, Global Rivers Environmental Education Network (OzGREEN). Retrieved from https://en.unesco.org/news/youth-take-pulse-planet-and-create-change-their-own-communities?fbclid=IwAR3TxDbh4mkV3SQ7VhqXRlH5YGMPKuaprV_Gu-JRAEut_9jpSidnprwKJ48

Main, G. (2005). *Heartland: The regeneration of the rural place*. Sydney, Australia: University of New South.

Masinire, A., Maringe, F., & Nkambule, T. (2014). Education for rural development: Embedding rural dimensions in initial teacher education preparation. *Perspectives in Education, 32*(3), 146–158.

McKenzie, P., Weldon, P., Rowley, G., Murphy, M., & McMillan, J. (2014). *Staff in Australia's schools 2013: Main report on the survey*. Retrieved from https://research.acer.edu.au/tll_misc/20/

Measham, T. G. (2006). Learning about environments: The significance of primal landscapes. *Journal of Environmental Management, 38*(3), 1–14.

Moll, L. C., Amanti, C., Neff, D., & Gonzalez, N. (1992). Funds of knowledge for teaching: Using a qualitative approach to connect homes and classrooms: Theory into practice. *Qualitative Issues in Educational Research, 31*(2), 132–141.

Moore, R. (1986). *Childhood's domain: Play and place in child development*. London, UK: Croom Helm Publishers.

Pettifer, L. (2019). Making sustainability happen: Activating environmental citizens and behaviour change in schools. *The Social Educator, 37*(2), 14–28.

Rasmussen, K. (2004). Places for children – Children's places. *Childhood, 11*(2), 155–173.

Reid, A., Green, B., Hastings, W., Cooper, M., & White, S. (2010). Regenerating rural and social space: Teacher education for rural-regional sustainability. *Australian Journal of Education, 54*(3), 262–276.

Reucassal, C. (2017). *War on waste* [Television series]. Sydney: Australian Broadcasting Company.

Ruston, S. (2008). *Activate your students: An inquiry-based learning approach to sustainability*. Carlton, VIC: Curriculum Corporation.

Thomas, S. (2011). Teachers and public engagement. *Discourse, 32*(2), 371–382.

Ulmer, J. (2016). Re-framing teacher evaluation discourse in the media. *Discourse, 37*(1), 43–55.

UNESCO. (2018). Youth take the pulse of the planet and create change in their own communities. Retrieved from https://en.unesco.org/news/youth-take-pulse-planet-and-create-change-their-own-communities

van Crowder, L., Lindley, W., Bruening, T., & Doron, N. 1(998). Agriculture education for sustainable rural development: Challenges for developing countries in the 21st century. *The Journal of Agricultural Education and Extension, 5*(2), 71–84.

12

HOPE FOR THE FUTURE

Enacting power and agency in ESE

Box 12.1 Extract from Daniel in his reflective response

To be honest, when trying to make small changes [in school practices], I thought more people would be like minded around sustainability – which on the surface they seemed to be. I realised that it had to be easy though and require little to no extra work load. In an already crowded curriculum with all areas fighting for time, if it was perceived hard or not a directive from the boss, then it would struggle to get traction.

(Daniel, reflective response)

This chapter opens with a reflective piece written by one of the early career teachers at the end of this project. By drawing upon this teacher's concluding remarks and summarising the key themes generated by this empirical study, we explore hopeful ways forward as we consider how teachers and learners alike can claim agency and power in furthering Environmental and Sustainability Education (ESE) in a range of educational settings. This chapter highlights 'the lessons learnt' to not only imagine new and innovative ways forward for agentive educators but for the whole community (including families, community and educational leaders, policy writers) to see teachers as empowered agents of change who are central drivers of ESE practices. Finally, we reflect on the contributions of this study which are two-fold: First, we argue that this book extends theoretical understandings of ESE implementation by exploring the power relations and structures within educational settings and the ways in which teachers attempt to balance their confluence of identities. Second, we claim that the use of a combinational methodology allowed for a non-linear, organic and interactive approach to capturing the experiential and emotion-laden stories of early career teachers that are frequently hidden from view.

In many ways, Daniel's story, seen above, epitomises the experiences of all the participating early career teachers in this project. Daniel's story illustrates that pivotal moment of realisation when the idealism they each took into their educational setting as an early career teacher was disrupted, rejected and/or misplaced within the reality of the educational system they had just joined. But their stories, and therefore this story, did not end there. Through the creation of a community of practice, the feelings, visions and disappointments of the early career teachers were shared and made visible. Their stories were looked at and examined, first in the workshops together, then during the time in-between by the researchers and the participants and brought back to the following workshop to look at collectively again. Over time, what the early career teachers thought they knew about being teachers, about being early career teachers and about being early career teachers who wanted to include ESE into their classrooms morphed into something different than what they had first envisaged.

This book concludes by summarising the two key themes, which were generated through the stories of the 11 early career teachers and the responses of the participating community partners, the community leaders and the researchers. First, the stories and responses highlight the ways in which early career teachers negotiated their sense of self, particularly in regard to their ways of thinking and doing (e.g. habitus) and how they balance their confluence of identities when this is seemingly disrupted. This negotiation, while at times led to disappointment and frustration, provided opportunities for sharing which then led to claiming back power and making small agentic steps forward. Second, this book reveals the power structures involved in educational contexts and how this influences teachers' habitus and creates tensions as these teachers attempt to navigate policies and curricula in a way that they can reclaim agency.

The (re)negotiation of self (habitus and confluence of identities)

At the start of this project, we aimed to explore 'the role that personal, professional and environmental identities of early career teachers play on the implementation of ESE in their everyday teaching practices.' Through listening to and analysing the experiential stories of early career over the course of six iterative collaborative workshops, we were able to identify the temporal, societal and place-based dimensions that influenced how they viewed themselves and the ways in which they attempted to align their habitus within the field, but also balance their confluence of identities (personal, professional and environmental).

The findings from this study, as realised within the stories of the participating teachers, reveal that these teachers acknowledged their passions and convictions about ESE and attempted to implement their ways of thinking into teaching practices. However, the study reveals that the majority of teachers, particularly Waruni, Kath and Jessica, had to significantly adjust their expectations of what they could do within their particular educational context, and celebrate 'baby steps.' Additionally, the participating early career teachers acknowledged the temporal dimension of implementing ESE and realised that this can take time and a keen sense of one's context. They realised that in order to implement ESE in their school or centre in a way that resonated with their sense of self, they needed to start small, creating a buzz and establishing buy-in from school/centre leadership, colleagues and learners.

This study also revealed how the teachers' habitus shifted to a more collective habitus—'my' identity to 'group' identity—both within the confines of their school/centre context and in the community of practice workshops. The teachers actively attempted to position themselves within the institutional habitus of their school/centre, being shaped by and shaping the school/centre's current ways of thinking and doing ESE. Some teachers, such as Shaun, Alison and Bonnie, relished being part of a school/centre community which promoted and valued ESE, while other teachers had to find ways to first understand the school/centre's habitus and the societal factors which influence it, and then navigate how to embed ESE in an educational setting that did not prioritise it. Given that a school/centre's institutional habitus is not developed in isolation but is reflective of their communities (Bartlett, McDonald & Pini, 2015; Brickson, 2007), the participating teachers began to realise that implementing ESE is not a simple and individual decision but requires the understanding of the broader context. Despite the different educational contexts and their varying views about ESE, the participating teachers realised that they were an integral part of these school/centre communities and that they must learn to harness the collective habitus of their institutions and the communities in which they were situated to know how to further ESE in their classrooms and their school/centres.

Additionally, the early career teachers embraced the power of being with like-minded colleagues within the community of practice workshops. These workshops provided a symbolically and physically safe 'third space' (Bhabha, 1994) for the early career teachers to share their emotionally charged stories and frequently marginalised 'first year' experiences. While unfortunate that these participants needed a designated 'safe' place to express their emotions about their ESE experiences, we realised quickly into the study it was a vitally important aspect that the early career teachers were requesting. Through their reflective and reflexive self-questioning during and in-between the workshops, the early career teachers were able to articulate similar concerns and problems whilst drawing upon a collective passion and desire to solve these problems (Jorgensen, Edwards & Ipsen, 2018). So too, these workshops enabled the teachers to regain hope, recharge and draw upon the energy of a collective habitus that shared the same values. Their collective problems resulted in collective solutions, which allowed them to feel they were not alone in their quest to work collectively towards ESE in their classrooms. In this way, the participating teachers were able to step beyond the dominant discourse of 'early career teachers' as 'incapable' and 'passively waiting during their first year of teaching' to acknowledge that they do in fact have valuable ways of knowing and acting as a teacher who wants to embed ESE into their curriculum.

The influential role of social contexts

As this project aimed to examine how teachers' habitus, and more specifically their confluence of identities, influenced their ESE practices, it became apparent that these early career teachers' ways of thinking and doing were shaped by, yet also shaped, the educational contexts in which they were situated, as discussed above. In answering our research question 'How do early career teachers negotiate their personal, professional and environmental identities in the midst of institutional habitus?', we found teachers did actively, over time (temporal dimensions), attempt to identify the institutional and societal norms that shaped their educational setting (societal dimension), while also utilising the place-based opportunities that empowered them to implement ESE within the context of their particular educational settings.

The findings suggest that powerful societal discourses, such as the media, shape educational contexts' approaches to the inclusion of ESE (see Chapter 11). While historically it is acknowledged that policies and curricula impact on how educational contexts view and understand ESE (Barnes, Moore & Almeida, 2019; Moore, Almeida & Barnes, 2018), this does not always result in a change in practices within educational contexts (Almeida, 2015; Munoz-Pedreros, 2014). This book suggests that educational settings interpret and translate these policies (such as curriculum) in a way that aligns with their institutional habitus, or their ethos, which, in turn, determines how they position ESE within their institutions (Bosevka & Kriewaldt, 2020). Therefore, ESE uptake seemingly is predetermined by the social or school context. While this provides a gloomy picture, the findings of this project suggest that early career teachers can successfully navigate these contexts so that they can enact ESE within the classrooms. With the support of the participating community partners, the participating teachers were encouraged to identify institutional norms and find ways to make ESE attractive to the priorities and ethos of their educational context. Some chose to start small, creating a buzz within the classrooms (Chapters 8 and 10) or by empowering learners to be social agents of change within the school/centre (Chapter 11), while others found a way to embed ESE into interdisciplinary planning documents (Chapter 9) or interdisciplinary inquiry-based learning units (Chapter 7). Despite these early career teachers' different contexts, they actively attempted to understand the institutional habitus of their school/centre and make agentic steps forward to embed ESE in teaching and learning. These teachers reclaimed power while simultaneously gaining more confidence, as they began to understand how to make meaningful change in a way that appeared to align with the social context in which they were situated. Additionally, many of the teachers, such as Eddie and Waruni, with guidance from the community partners (Chapter 6), came to the realisation that institutional change occurs when learners are empowered and enabled to take the lead (Davis, 2010).

The findings of this project also reveal the importance of place-based learning and the opportunities it provides for connecting and engaging with the local community (Allen-Gill, Walker, Thomas, Shevory, & Elan, 2005; Chan, Mathews & Li, 2018; DePetris & Eames, 2017; Moore, O'Leary, Sinnott & O'Connor, 2019). The teachers in this study identified place-based opportunities (whether this was in outside spaces or connecting with community partners) that could create meaningful opportunities for teachers and learners alike to connect ESE with the local spaces around them. Despite some educational contexts not fostering opportunities to engage in ESE, the teachers became aware that they were not alone and there were spaces that allowed for ESE engagement which connected them with other like-minded people within the community. These 'spaces' came in the form of the community of practice workshops, the place-based opportunities that were provided through the community partners, Centre for Education and Research in Environmental Strategies (CERES) and Dolphin Research Institute (DRI) and other community leaders, such as the Melbourne Zoo and Birdlife Australia. Many of these organisations had existing spaces available for schools to interact and engage with ESE but they also provided materials, resources and camaraderie for the teachers who had felt alone in their school/centres.

Lessons learnt from this project

The stories of the participating early career teachers provide hope through the suggestion that teachers *can* enact agency in ESE implementation; however:

1. This takes time and patience;
2. It requires an understanding of the institutional habitus of one's educational context so as to identify ways to embed ESE in a way that resonates with both one's individual habitus and their school/centre's institutional habitus. They can actively find ways to make ESE meaningful for their particular contexts…even if this means starting small;
3. Teachers' identities need to be acknowledged with opportunities to strengthen these;
4. Teachers need to be recharged and empowered by connecting, sharing and problem-solving with others who are passionate (even a *little bit* passionate, as Jessica suggested) about sustainability and the enactment of ESE;
5. The local community needs to be utilised as there are endless opportunities to connect with like-minded people and embrace local spaces as a resource for ESE;
6. Learners (including young children) need to be positioned (and/or re-positioned) as the actors or agents of change, which will create a buzz that is difficult for schools, centres, parents and politicians to ignore.

We argue that this book contributes to the intersection between the fields of teacher education and ESE. Theoretically, we argue that to critically examine the enactment of ESE, one must theorise teachers' confluence of identities (personal, professional and environmental) and the power structures and relations that exist within educational contexts. This research shows that while individual identities are crucial, it is equally important to find ways for strengthening these towards a confluence that is larger than the sum of individual parts. Communities of practice offer those opportunities where personal and environmental identities bolster professional identities allowing a robust implementation of ESE in everyday practices. While exploring the power relations within education is nothing new to educational research, its use alongside the idea of powerful knowledge helps in understanding how power is exercised not only through important social actors but also how it is exercised through identifying 'knowledge' that is valued and prioritised. The ways in which ESE is positioned as an (un)important knowledge area as well as how early career teachers are positioned as legitimate contributors of ideas and practices in their new school/centres are important so that teachers can respond to these power structures with power. Ultimately, in understanding the positioning of ESE as well as the precarious positioning of early career teachers, educational contexts can be navigated in ways that reclaim power.

Additionally, we suggest that a combinational methodological approach, which employs narrative inquiry and research by design, provides a rich and in-depth view of the participants' experiences. The individual methodological approaches used in this project are not new to educational research. However, we argue our methodological contribution is through the purposeful combination of a research by design framework with the principles, practices and analytical processes of a narrative inquiry embedded within it as the overall methodological approach. Through the combination of these complementary, intrinsically organic, methodological approaches, much stronger procedural and analytical processes were created for this project than would have been possible had only one of the individual methodologies been used. The research by design processes provided the iterative framework with a series of modified and adapted 'interventions' co-constructed between the participants, community partners and researchers running throughout the project. Within this framework, the narrative inquiry enabled the rich source of conversational interviews which triggered and created the storied data. It was this storied data which was then analysed through the three-dimensional narrative inquiry model of examining the temporal, societal and place-based dimensions, leading to the identification of findings that formed the basis of each of the narrative chapters in this book.

If, as the substantive literature and global and national policies claim, education is the decisive factor in the success of ESE (cf: UNESCO, 2015), then the findings from this project will assist in reframing the way early career teachers are viewed as a significant cohort within educational systems. Rather than underestimating, negating and dismissing their roles and identities, early career teachers can now be seen to be valuable educators highly capable of contributing ideas, ideals and innovative approaches to the common causes in ESE while working and teaching alongside other teachers, learners and the community towards a hopeful future.

References

Allen-Gill, S., Walker, L., Thomas, G., Shevory, T., & Elan, S. (2005). Forming a community partnership to enhance education in sustainability. *International Journal of Sustainability in Higher Education, 6*(4), 392–402.

Almeida, S. C (2015). *Environmental education in a climate of reform.* Rotterdam, The Netherlands: Sense Publishers.

Barnes, M., Moore, D., & Almeida, S. (2019). Sustainability in Australian schools: A cross-curriculum priority? Prospects, *47*(4), 377–392. doi:10.1007/s11125-018-9437-x

Bartlett, J., McDonald, P., & Pini, B. (2015). Identity orientation and stakeholder engagement: The corporatisation of elite schools. *Journal of Public Affairs, 15*(2), 201–209.

Bhabha, H. K. (1994). *The location of culture.* Abingdon, Oxon: Routledge.

Bosevska, J., & Kriewaldt, J. (2020) Fostering a whole-school approach to sustainability: Learning from one school's journey towards sustainable education. *International Research in Geographical and Environmental Education, 29*(1), 55–73. doi:10.1080/10382046.2019.1661127

Brickson, S. (2007). Organizational identity orientation: The genesis of the role of the firm and distinct forms of social value. *The Academy of Management Review, 32*(3), 864–888.

Chan, Y-W., Mathews, N., & Li, F. (2018). Environmental education in nature reserve areas in southwestern China: What do we learn from Caohai? *Applied Environmental Education & Communication, 17*(2), 174–185. doi:10.1080/1533015X.2017.1388198

DePetris, T., & Eames, C. (2017). A collaborative community education model: Developing effective school-community partnerships. *Australian Journal of Environmental Education, 33*(3), 171–188. doi:10.1017/aee.2017.26

Davis, J. (2010). Chapter 1. What is early childhood education for sustainability. In J. Davis (Ed.), *Young children and the environment: Early education for sustainability* (pp. 21–42). Cambridge: Cambridge University Press.

Jørgensen, R., Edwards, K., & Ipsen, C. (2018). Intentional development of communities of practice: Improving knowledge sharing and work guidelines. *European Conference on Knowledge Management.* Implementation and performance management, DTU Management. Academic Conference & Publishing International Ltd., Denmark.

Moore, D., Almeida, S. C., & Barnes, M. (2018). Education for sustainability policies: Ramifications for practice. *Australian Journal of Teacher Education, 43*(11), 105–121. Retrieved from https://ro.ecu.edu.au/ajte/vol43/iss11/6

Moore, M., O'Leary, P., Sinnott, D., & O'Connor, J. (2019). Extending communities of practice: A partnership model for sustainable schools. *Environment, Development, and Sustainability, 21,* 1745–1762. doi:10.1007/s10668-018-0101-7

Munoz-Pedreros, A. (2014). Environmental education in Chile: A pending task. *Ambiente & Sociedade, 17*(3), 175–194.

UNESCO. (2015). Sustainable development goals – Education. Retrieved October, 2020 from https://sdgs.un.org/topics/education

INDEX

For Product Safety Concerns and Information please contact our EU
representative GPSR@taylorandfrancis.com
Taylor & Francis Verlag GmbH, Kaufingerstraße 24, 80331 München, Germany